D1488237

A NEW AMERICAN SPACE PLAN

= BY =

TRAVIS S. TAYLOR

Ringleader of the
ROCKET CITY REDNECKS

A NEW AMERICAN SPACE PLAN

BY

TRAVIS S. TAYLOR

Ringleader of the
ROCKET CITY REDNECKS
The National Hit Television Series
with Stephanie Osborn

BAEN

A New American Space Plan by Travis Taylor,
Ringleader of the Rocket City Rednecks

A Baen Books Original

Baen Publishing Enterprises
P.O. Box 1403
Riverdale, NY 10471
www.baen.com

ISBN: 978-1-4516-3865-3

Cover photo © 2012, Donnie Claxton

All interior photos © 2012, National Geographic Channel. Used by permission.

First Baen printing, November 2012

Distributed by Simon & Schuster
1230 Avenue of the Americas
New York, NY 10020

Printed in the United States of America

10 9 8 7 6 5 4 3 2 1

A NEW AMERICAN SPACE PLAN

BY

TRAVIS S. TAYLOR

Ringleader of the
ROCKET CITY REDNECKS

★ ★ ★

with Stephanie Osborn

CHAPTER I:
WHERE IS CAPTAIN KIRK
WHEN WE NEED HIM?

When I was a kid I watched Captain Kirk every afternoon when I'd get home from school. On weekends we pretended to be Captain Kirk; there was always an argument on who got to be whom. I always wanted to be Kirk or Spock but everybody else did too. Every now and then I pretended to be Mr. Scott, but James T. Kirk was the one we all truly wanted to be.

In our backyard there was this huge pine tree. That thing had to be at least a hundred and twenty feet tall and was so big that my older brother and I could not reach around and touch each other's hands on the other side. We built a platform up in the top of that tree and we would climb up that thing all the time. We didn't build that platform just to make our parents nervous or just to have a treehouse way away from everybody. No, that wasn't it at all. You see we were just outside the Rocket City—Huntsville, Alabama. When we climbed up the tree we could see everything in north Alabama, including across the river where the NASA Marshall Space Flight Center was. It was right there on the Redstone Arsenal where Wernher Von Braun started the whole space thing. When they would do engine tests across the river, we could see the smoke plumes. We could hear the roaring thunderous sound of the rocket engines pouring out exhaust and thrust, and it would rattle the windows of our house.

Man I miss those days. I miss that excitement. That roar into the unknown. I miss that feeling that we were going to go out there to the unknown and get to know it, tame it, and make it ours! That is what America is missing right now.

I remember being thrilled, just absolutely thrilled, to get to watch the rocket testing. It made me really feel like Captain Kirk. At one point, when I was about ten or twelve or so, NASA flew the new Space Shuttle Enterprise test vehicle into Huntsville, Alabama. It was parked out at the Space Center and I went and took pictures of that thing.

There was no doubt in our minds that we were going to get to retire on the moon. And our kids no doubt would be able to live up there or at least vacation there. And their kids were going to get to vacation on Mars. There was always the hope that somebody in those big government research programs was going to end up with a warp drive and we would soon be exploring Rigel 7, or Tau Ceti 8, or even Vulcan.

Kids growing up today have no clue how absolutely thrilling and exciting it was to watch Captain Kirk and Mr. Spock on the TV in the evenings. And to feel it and see it. See it—in the form of the current space program—on the news in the afternoons. And just to know, by God, that America was first in space, would always be first in space, and that one day all of us would have the opportunity to *go* to space. That final frontier was ripe for the picking and it was America's choice to pick. We were all going to be Captain Kirk!

So what the crap happened!?

When I was in high school everybody wanted to be a scientist, an engineer, or a fighter pilot so they could become an astronaut. If they didn't actually fly in the things, they at least wanted to work on 'em. Spaceships that America was building to take us to the final frontier were going to be a reality very soon, seemed like. When I was in graduate school there were one hundred and twenty students in the physics department at our local university. There were hundreds of kids and we had all the engineering curriculums. Now twenty years later there's about eight kids in the physics department. But at least the engineering department is still alive and kicking. Mechanical and aerospace do okay because of the big need for UAVs nowadays. But there is almost zero work being done on space vehicles.

I was peripherally on a program that NASA tried to do for about two or three years called "Breakthrough Propulsion Physics." One bright guy at NASA named Mark Millis hoped that he could use enthusiasts, tenured professors, and graybeards across the country to, on their own dime mostly, work on concepts like warp drives and teleporters. It didn't last long because there was no money in there. I think throughout the total length of the program there were a few million dollars to be had for government, academia, and industry. Think about that. After overhead and taxes, a single person, whether he makes $50,000 or $100,000 per year, still costs the company, government, or university about $250,000 per year. So, the Breakthrough Propulsion Physics program had enough money to keep about five people employed per year. Two of them were at NASA, Mark Millis and his assistant. So, that left enough money to keep about three more people thinking about how to build a warp drive. The most difficult breakthrough physics in the history of mankind is not going to be accomplished by hiring three people.

But Mark was hopeful that we'd all jump in there and start to work on these problems together even if there was no money. The thought was that if enough enthusiasm picked up maybe we could lobby for more money from Congress. The reality of the situation was that the only people who ended up able to work on the effort were the tenured professors, a few enthusiasts like myself, and that's about it. With no real money to keep people interested, the program died out. There was not enough enthusiasm from the general public to bolster and shore up the next generation in our American space program.

We had had a good start. Back when I was a kid, everybody knew the names of the Mercury Seven, the New Nine—who were the Gemini team—and every Apollo astronaut. Then there was a long hiatus of almost ten years before the Space Shuttle ever launched, and once it had, some of the enthusiasm came back. Most of the kids my age knew who John Young and Robert Crippen, Jr. were. They weren't quite as strong as Buzz Aldrin and Michael Collins, let alone Neil Armstrong, but they were the new generation of astronauts. Of course John Young was famous from the previous era and we all knew that he was perfect to be the first commander of the International Space Station. After all, he flew on the first manned Gemini mission, Gemini 3, with Gus Grissom; he was the first man to orbit the Moon alone in Apollo 10; he commanded Apollo 16, walked on the Moon, drove the lunar rover, and *still* commanded STS-1—the first manned Shuttle mission—*and* STS-9, which carried the first Spacelab payload. First, first, first.

Who could ever forget about John Young's corned beef sandwich? How on earth he ever smuggled it onto Gemini 3 I can't figure. It got him in trouble—he got reprimanded for it—but it was still another space first to add to an impressive list.

What do we do now? How do we get the fire back in our collective national gut? I don't know what we can do to bring that enthusiasm back. When Steve Jobs looked into where the new iPhone would be constructed, the study showed that it would take ten years or more for Americans to train enough engineers and scientists and technicians to manufacture his cool gadgets. But China could do it *right now*. Guess where he went?

NASA hasn't been able to hire new folks for a while, ever since the Constellation program went away. The current presidential administration has killed tens of thousands of jobs in the space industry and we've lost those scientists and engineers and their capabilities. I mean *lost*. They had to pay the bills so they went—and took jobs elsewhere, doing other things, and the space program has lost that legacy knowledge. And lost that enthusiasm for what could be done next. With all the starts and stops in the space program since Shuttle, NASA morale is like that of an abused child—continuously battered and then praised from the same people, and then battered again. NASA has become overly cautious to do anything and that includes being enthusiastic about any new program because you never know when that program will be yanked out from under you.

But it didn't start with this administration. I may or may not agree with anything that the current administration has done, but I can't blame it completely for what is happening to our space program. So who's to blame? Why have we let our lead in space slip away? Why is it that our once solid grasp on space is now almost all but gone? What happened?

Politics.

The way it works is that our space program tends to change in flavor, style, shape, fashion, and mission plans based on every election. If a previous administration had a bold plan and it was moving forward, typically the next administration kills it just for political spite. They want to put their own stamp on things. The hiatus between Apollo and Shuttle was clearly politics because President Nixon wasn't about to allow one of the most successful efforts in his administration's tenure in office to be the brainchild of the Kennedy administration. The change occurred when a different political party came into the White House . . . right after the Apollo-Soyuz detente flight. Politics. Then we didn't fly for a decade or so. It took that long to recover from politics.

I was about to start into junior high school and *Star Trek* had been in reruns for almost a decade. Our childhood hopes of space were slowed but not dead. The Apollo era wasn't too far behind us and we all still believed we were destined for space. Captain Kirk was still out there somewhere with rumors of there being a new *Star Trek* movie about to happen. Luke Skywalker, Han Solo, and Princess Leia were getting our attention. Space hadn't been killed completely. Not yet. But, in reality, we had stopped going to the Moon and hopes for Mars were so far away as to be almost laughable when seriously talked about. The thought was typically, "Mars, hah! Not in my lifetime."

Captain Kirk, we truly, truly, need you now.

CHAPTER 2:
THE INTERNATIONAL
SPACE STATION

When we started out making plans for the first season of *Rocket City Rednecks* I really wanted to build a spacecraft—an actual vehicle that flew into space and performed some task. Well, the budget for such a mission just wasn't available. So, I thought hard about what might be as difficult and fun and entertaining to boot, as well as a build that we could learn a lot from. In the end, I decided on building a submarine.

None of the five of us "rednecks" (Daddy, Pete, Rog, Michael or I) had ever built a submersible vehicle before but to us, that doesn't mean we couldn't do it. It just means it is a big opportunity to learn how. Originally, I wanted to build a sub large enough to accommodate all five of us and something we'd keep around and use after the show. We brainstormed how to do such a thing. I thought about concrete, steel, fiberglass, and several other building materials. The problem with using those types of materials meant that we would be building everything from scratch. Now, keep in mind that all of our builds are scheduled to take only one weekend. There was no way we could build a five man submarine from scratch in one weekend. There was just no way.

Then, on the way home from work one afternoon I saw a polyethylene plastic tank sitting out in the edge of a cotton field and I had the idea. A plastic tank would work. The farmer's tanks are rugged and watertight. The only issue was that they were designed to hold water inside, not keep it from *getting* inside. That meant we had to develop some structure to keep the walls from caving in on us. It turned out that this part of the design would be very important later on during our final test, but I'm moving ahead of myself.

The plastic tanks just so happen to come on pallets with steel reinforcement cages on the outside. We bought two tanks and left one cage outside and then cut the other down and used it for reinforcement on the inside. This wasn't our original plan, but once we decided that two of us could fit inside one tank we had the extra just sitting about. Then my buddy Rog and I, actually on camera, had the idea at the same time to use the one cage from the other to go inside the sub to support the plastic tank from the inside.

We worked really hard mounting the tank to an old boat trailer of Daddy's and then welding up ballast tanks. Our ballast tanks were stainless steel beer kegs with the bottoms cut out and an air hose fitting welded in the top. We could blow air into the keg and it would force water out the bottom and therefore increase the buoyancy of the sub. We added three of these to the sub design.

After two days of work on the sub we decided that we needed to test it. What you didn't see if you watched the episode was that we took it down to the Tennessee River. My nephew Michael and I drove it around for a good while above surface in the river and it all worked very well. The big and really only issue was that the carbon dioxide level went through the roof in just about fifteen minutes. We actually hit fairly dangerous levels extremely quickly. That meant that we needed to build a CO_2 scrubber to remove the toxic gas from the cabin. I then did a calculation and figured out that the CO_2 would be dangerous in about twelve minutes. Fortunately for us, we had planned ahead and had oxygen in the sub with us for the test. Once our CO_2 meter started beeping we got on the air and stayed on it. But I didn't want to do that all the time while in the sub, so, we decided to build scrubbers.

Apollo 13 had some serious issues pop up with the CO_2 scrubber, and the astronauts actually had to repair it while on their way to the moon. With hindsight, I wish we'd just left the CO_2 scrubber out of the build and stayed on a rebreather the whole time. That would have been less convenient, but the end result with the scrubber was even more inconvenient. But that is a show spoiler I'd hate to give away. Well . . .

If you insist . . .

The sub worked great at a depth of twelve feet. We stayed under for more than a half hour with no problems. Once we decided to go to sixteen foot is when we had an issue. We had built the scrubbers out of two five-gallon plastic buckets with lids. We put sofnolime material in the bucket and forced our air through those buckets from within the cabin, through PVC pipe, and then into the bucket through the sofnolime, which scrubbed out the CO2. Then the air went through another pipe and into a second bucket of sofnolime and then back into the cabin. It had been working all right. But at sixteen foot the seals on the plastic five gallon buckets gave way and they instantly filled with water.

The missing air volume of the buckets made us less buoyant and we sank like a rock right to the bottom. The walls started caving in and buckling and seals started popping! I told my nephew to add pressure to the cabin and that worked. Temporarily.

The pressure held the structure of the cabin intact but it also forced air through the scrubbers, which was at this point a sofnolime and water-chemical mix. Some of this mix was calcium hydroxide gas—not something you want to breathe or even just hang out in. The toxic gas was forced into the cabin as a milky-white, skin-irritating, eye burning, lung searing fog.

At this point, it was time to evacuate. Fortunately, over-pressurizing the cabin had made us overly buoyant and we popped to the top like an Ohio-class submarine. We set to work undoing the hatch to get out. What we didn't realize was that we had five full grown safety divers on top of us attempting to open the hatch from the outside. We hadn't designed the sub for surfacing with that much weight. Needless to say, when I opened the hatch from the inside we were three inches below the surface of the water. The cabin instantly filled with about 350 gallons of water thus taking us all the way to the bottom at about twenty-two foot. We could do nothing but hang on for the ride. Once at the bottom we swam out and held our breath to the surface. It was a long twenty-two foot.

So, what does this have to do with the International Space Station? Mainly, our sub was a lesson in doing things we don't know how to do yet. That is what building the ISS is all about. We are learning how to do construction in space which is something we have never done before. By doing it we are learning things that are important like, don't use 5 gallon buckets to put your CO2 scrubbers in.

Like anything, the ISS has its warts and ugly spots just like *our* submarine. But, in the end, it was our submarine and it is *our* space station, ugly spots and all. Honestly, I can't wait to build another submarine and I most certainly would go to the ISS in a heartbeat.

★ ★ ★

So we quit going to the Moon by the time I was three or four years old. Then we built the Space Shuttle to hold my adolescent and young adult interests. But that would only take us to low Earth orbit, or LEO for short. In all the dog-gone years since Apollo, we have not gone past low Earth orbit. We spent all of our space investment money on the International Space Station. Let me tell you, there's nothing *international* about it. There's a color-coded picture in the Augustine committee report on America's space program that shows yellow as American components and other colors from other countries. And, lo and behold, about ninety percent of that thing is yellow. And what we don't tell other people, or admit to, is that we paid a lot of the countries to build those other pieces that don't code in yellow. International my butt! Oh, there was international cooperation and there were components and investments from other countries, but there would have been no station without America.

We also let politics dictate the orbital inclination we put the station in. It is up around fifty-one degrees so it could pass over the Russian launch site. An optimal orbit for an orbital ship building factory would be closer to the equator. But we'll get to this part later.

Even though the space station is in a nonoptimal orbit to be a shipbuilding platform, it is still an amazing vehicle. It has been in space longer than any other manned spacecraft. It's an amazingly large spacecraft. If it were placed in a college football stadium it would reach from end zone to end zone, from sideline to sideline, and would stretch all the way up into the second seating levels. It can dock with multiple spacecraft from multiple countries. And it even has a back porch.

So, no matter the orbit, the International Space Station, or ISS, *is* an incredible habitat. It has been in orbit for about twelve years now. During that time, it has been constantly growing and developing. As of this writing, it has a module length of 167.3 feet (51m) and a mass of 861,804 lb (390,908kg), the equivalent of over three hundred cars. The truss measures 357.5 feet (109m) long and the eight solar arrays are some 239.4 feet (73m) long, generating on average 84kW of power. It is a small city in space.

I'm not truly concerned that the ISS is in the wrong orbit. Oh, I wish it were in a better one, but we can live with it where it is. We understand orbits really well and have docked with the space station many times even though it's at the high inclination. It just requires a little more fuel to do the inclination cranking maneuver. We could still use it as a shipbuilding yard. The only issue would be whether we decide to launch to the moon or Mars (or wherever else we were going), we would have to crank down the inclination of our inter-planetary vehicle. The other option would be to go ahead and move the space station as discussed previously. In the end, this means doing a little extra orbit calculations and maneuvers and spending money on a little more fuel. The saving grace the ISS has is that it is just so large and it *is* in space already. We shouldn't forget that. The ISS *is* in space already and we paid a lot of money to *get* it there. We should *keep* it there.

It has a habitable volume of 13,696 cubic feet (388m³) but a pressurized volume of 32,333 cubic feet (916m³). In its twelve years of operation, that habitable volume has seen more than two hundred astronauts and cosmonauts living within it. It is almost four times as large as Mir was, and five times as large as Skylab was. Its living space is bigger than a standard five-bedroom house. It has two bathrooms, a gymnasium, and a three-hundred-degree bay window, for a panoramic view of the heavens unequalled on Earth.

As of its tenth anniversary in 2010, it had made 57,361 orbits of Earth, covering over 1.5 billion statute miles (~2.5 billion km), or 16 astronomical units. Five different types of launch vehicles were used in its construction (the U.S. Space Shuttle, the Russian Soyuz and Proton, ESA's Ariane 5, and Japan's HII[1]), and fifteen nations substantially participated: the United States, Canada, Japan, Russia, Brazil, and ten nations of the European Space Agency, including Germany, France,

---

[1] http://www.nasa.gov/externalflash/ISSRG/pdfs/launchvehicles.pdf

Italy, Belgium, Switzerland, Spain, Denmark, the Netherlands, Norway, and Sweden.

More than 160 EVAs have been conducted in its construction, totaling over 1,000 hours in space. It requires some 2.3 million lines of computer code onboard, and an additional 3.3 million lines on the ground to operate it, with fifty-two onboard computers. It requires eight miles of electical wiring. It draws 75 to 90 kilowatts of power, provided by an acre of solar panels. By contrast, the average home draws about one and a half kilowatts of power.[2]

The fifty-five-foot-long robotic arm is capable of lifting a Space Shuttle.

Its orbit is 250 miles (402 km) high, with an inclination to the equator of 51.6°. This orbit not only allows for easy access by all launch vehicles of the international partners, but it also enables viewing of eighty-five percent of Earth (and ninety-five percent of the world's population) for scientific endeavors. If the ISS is implemented more for Earth science then the orbit isn't so bad.[3]

What good does a space station in low Earth orbit do us? For a start, it's a close outpost. We can get there in a couple hours, and we can do lots of experimentation and training there that will allow us to go farther, and to learn how to do some things that will be beneficial, not just in space, but on *Earth*. Already there's been some fascinating research on the ISS.

We've been doing science experiments on the ISS pretty much since the first module was put in orbit. ISS onboard research has included protein crystal studies, tissue culture, low gravity medical and biological effects, metallurgy, flame research, fluid research, space environmental studies, and Earth observations of natural phenomena.

Since gravity drives convection, solutions in a microgravity environment like ISS tend to be much less disturbed by unwanted motion than they do on Earth. This enables us to grow much larger

---

[2] http://www.nfpa.org/assets/files/PDF/Research/PowerConsumption.pdf

[3] For more details on the ISS I suggest you check out the following Web sites:
   http://www.nasa.gov/mission_pages/station/main/onthestation/facts_and_figures.html
   http://www.shuttlepresskit.com/ISS_OVR/index.htm]
   http://en.wikipedia.org/wiki/ISS

crystals, because it takes an undisturbed environment to do so. The proteins being grown into crystals include those from bacteria, enzymes, and viruses. If the crystals are large enough to use X-ray crystallography to determine their chemical structure, then much can be learned about their chemistry and interactions that we couldn't learn on Earth. In turn, this may enable us to develop cures for some diseases, such as HIV, cancer, and diabetes.

Likewise, the absence of convection causes flames to behave differently on orbit. Instead of the usual teardrop shape we are used to with a candle, which is formed from air convection around the flame, flames in the microgravity environment form spheres initially, then toroids as they burn outward from the original ignition source. If an air current is introduced (as if from a vent), the burn becomes preferentially upwind. Not only may this help us in better understanding the behavior of wildfires on Earth, it may aid us in designing warning systems for fire and smoke detectors when we venture beyond LEO.

The lack of gravity enables a normally immiscible (unmixable) pairing of fluids (the classic being oil and water) to mix, forming different types of emulsions. Should these fluids be, for instance, molten metals, then it becomes possible to make alloys in space that could never be made on Earth. And who knows what kind of salad dressings we could come up with?

As I mentioned previously, the lack of gravity enables crystals to grow much larger than they might on Earth. If properly isolated from accidental movement, they can grow for days, weeks, even months. This enables medical studies of specific proteins. It seems that we do not know for certain the chemical structures of some vital proteins, such as those which determine the ratio of "good" to "bad" cholesterol in the bloodstream, or that activate viruses such as HIV. The ability to grow large crystals of these proteins enables us to determine their structures using such techniques as X-ray diffraction spectrometry.

Having definite structures may, in turn, enable us to develop drug treatments that activate this protein, or deactivate that one. Hopefully, we could use what we learned in space to turn on or off good or bad proteins as the situation warrants.

Medical and biological research tell us more about how life forms behave and how they react to certain stimuli. Performing them in the microgravity environment of ISS enables us to isolate them to specific stimuli. It also tells us much about our own bodies and how they react to space travel, and enables us to begin planning for a safe journey to another planet—whether that planet is in our own solar system or another one.

While building the ISS and working in, on, and outside it we have really learned a lot about space itself. Space environmental studies tell us more about the world outside our world. "Open space" is a harsh environment, filled with radiation of all types (particles and photons), micrometeoroids, extreme hot and cold temperatures (depending on whether you are in the sunlight or in shadow), and other hazards. Not all materials respond to these conditions well. Even on Earth, the extreme cold in the polar regions can cause even the hardest metals to become very brittle. Being able to stick a new alloy sample on the outside of your home to expose it to space can be an easy way of finding out how tough it really is. For reasons like this simple test we need to maintain the ISS as long as we possibly can. We need to add to it. We should put every type of foundry, factory, and experiment we can think of up there so we can have better things down here.

And space-based observations of Earth phenomena such as erupting volcanoes, hurricanes, or even simple thunderstorms can help ground-based scientists gather more data on the subjects of their studies, potentially enabling better forecasting and prediction. We've learned more about the high altitude peaks of thunderstorms and the lightning up there from the continuous view from the ISS than we ever did before. The continuity ISS offers is like no other scientific platform.

But, like my grandma used to always tell me, "every frog has warts." Unfortunately, despite all these wonderful things the ISS can do, there are a few flaws that should be pointed out. From what I've learned about it over the years working on or close to experiments that have flown up there and from what I've heard from some friends that worked the program closely, there are things the general public doesn't know. There are things that would not have happened if the original, U.S. Space Station Freedom concept had been maintained. Because of politics and not sound engineering we made some design flaws acceptable.

You see, when President Clinton announced a change in the direction of the station project, sending down an administrative directive that it would be an international, rather than strictly an American project, something got left out. An oversight group. Put bluntly, there was no single agency responsible for the overall design management of the construction of ISS. So when all of those international partners put together their segments of ISS, they did it to their own country's standard specifications. And anybody who has tried to take an American hair dryer to Europe and plug it in and get it to work knows exactly what that means.

It means that power and voltage vary from area to area inside. To flow power from one segment to another requires converters. It means that computers within the Station don't necessarily talk to each other, and can't be networked together unless a "sneakernet"—yes, literally walking (or in this case, floating) a disk or flash drive from one machine to another—is used. It means that different segments have their own independent "air to ground" communications systems, although there is an internal "telephone" system installed.

Early on, I understand, they even had problems with the solar arrays—there was a tendency to build up static plasma charges around the panels. Doesn't sound like much until somebody tells you that they had to power down lots of the Station before an EVA, shut off the solar arrays, and let the static charges dissipate before allowing an astronaut outside the hatch on a spacewalk. Why? They were afraid the arrays would arc and electrocute an astronaut. Think about it. That's one heckuva big static charge, millions of volts even. But it was supposed to get fixed in time. I'm not certain if that issue was ever solved.

Oh, and then there's the matter of payment. Just like with the launch vehicles, most of the modules ended up being paid for by the U.S. Not all, but way more than you'd expect, given an "international" station. One nation, who shall remain nameless, but who was operating their own space station, was actually reported to be using the funds NASA was providing (out of a limited NASA budget, mind you, about which we'll talk more on) to maintain their own aging station. There was also a hushed-up scandal about high-level space and military officials in that nation (often said officials were one and the same) lining their pockets with the funds and purchasing vacation mansions in resort areas of their country. The end result was that the module that was eventually provided had been paid for by the USA several times over by the time it was actually put in orbit and installed. This is all hearsay of course, but it is hearsay that is pretty much accepted as fact throughout the community.

Due to politics we can't really say it is "our space station" but the other countries sure can, and do. The payload flight controllers were actually instructed that, should the flight controllers or crew from this other nation we've been talking about refer to ISS as "THEIR station," our controllers were not to contradict them. It was a "cultural thing." And now this country is considering ditching "their station"— for reasons I don't really understand, because despite its problems, it is still a viable base of operations and scientific research—within a year or two, and moving on to something else. They are even pushing for deorbiting ISS and letting it burn up on reentry. But they do intend to separate "their" modules from the rest of the ISS before deorbit, to use for their own independent, internally built station.

All of these problems could have been alleviated if politics had not interfered and politicians had not insisted upon an international "exchange." Had ISS remained Space Station Freedom, it would be built to uniform specs throughout. It would have modules that communicated, shared power, and networked computers. There would be no need for converters or sneakernets or multiple air to ground links. There would have been a savings in cost by avoiding the diversion of funding through other nations. There would have, in systems engineering terms, been a Lead Systems Engineer, a Lead Systems Integrator, a Configuration Management Team, and one Program Manager.

There has also recently been some political talk within the Obama administration of letting the space station program die after a few more years—maybe five to ten or fewer years depending on budget. This would be a huge mistake. If we were to walk away from the space station the Russians would cannibalize it for their own Russian space station. Apparently, this is already in the plans that the Russians have. And at that point, what would keep the Chinese from flying up to the space station and cannibalizing it themselves, using salvage laws to support their claim?

Now our best hope would be to keep the space station alive, keep improving it, keep repairing, and keep moving forward with it. Build a module specifically designed for integration of an interplanetary space vehicle on it. We need a machine shop and shipbuilding yard in space. Having the platform in orbit with interplanetary transfer vehicles in place there would make traveling to and from the Moon, and in the future to Mars, much simpler. The initial cost might be larger, but in the long run it would work pretty much like the hub concept in most transportation systems, including the airlines, trains, and subway systems. The local airports would be the launch sites around the world. And the interplanetary hub would be at the International Space Station in low Earth orbit. Then of course we would need to create a hub in orbit around the Moon. And, using that hub at the moon, we would then transfer to local lunar sites and areas. And there would be no talk of deorbiting it now, when it was intended to operate until at least 2020, and possibly 2028. At the moment, however, we can't even get to the International Space Station ourselves without paying the Russians to give us a ride.

CHAPTER 3:
WE NEED TO PICK A WAY TO SPACE AND STICK WITH IT!

A typical season of *Rocket City Rednecks* starts out with the production company Flight 33, producers from National Geographic Channel, and myself sitting through several long teleconference calls where I pitch about thirty ideas for shows. We argue, discuss, debate, argue some more, and then eventually come to a consensus on what builds would work best for television. We usually pick tentative builds for all of the coming season's episodes. Every now and then we will swap out an episode with a new idea if one comes up that is better, but for the most part we know the general idea of what we want to build at the beginning of the season. Of course, I don't say I know at that time how in the world we would ever accomplish such a build. Most times I feel like Will Smith and Jeff Goldblum's characters from *Independence Day* must've when they explain that they can fly up to the alien spacecraft and implant a computer virus on it.

"Do you really think you can fly that thing?"

"Do you think you can do all that bulls@#t you just said?"

Well, I always feel that way up until Sunday after we have completed the build and tested the idea. A lot of times even afterwards, because we have a lot of stuff blow up on us. If you've seen the show then you know that sometimes we fail spectacularly. That's all part of science.

But the key thing here is that we pick a goal and stay with it.

Before we build anything we always do a little bit of planning. Sometimes it is required because the parts are hard to find or have to be ordered. Sometimes, it is because we have to use a certain place for testing or a particular thing or service for safety that is only available on certain days. But this is more of the logistics planning as opposed to actually having a "plan." The plan we Rocket City Rednecks use is typically to be stubborn as a mule. In fact, Rog has called me "mule" for years. Some might argue, including my mom, that I am more stubborn.

What we are stubborn about is that we say Friday morning that we are going to build a "such and such" and, by God, before the end of the weekend (or in two cases, the Junkyard Iron Man and the submarine by the end of the next weekend) we WILL have built a "such and such" and tested it at least once. Every now and then we might hit a point where we wish we could buy a certain widget, like with the Iron Man suit. I wanted to have more expensive actuators and servos and more time to write control software. We didn't have time or budget for that. After all, we only get a few thousand dollars per build. That is one reason we scrounge in junkyards, hoard parts at Rog's house (Mom won't let us do it at hers and neither will my wife), and use mostly prepackaged canned software programs. We just don't have time or budget to dig deeper. So, sometimes we compromise on performance, but never have we stopped in the middle and said, "Hey, let's stop this, throw all of this work in the trash, and do something completely different." That would be a waste of time, money, and a lot of elbow grease.

The problem here is that this is *EXACTLY* what NASA has done all my life. They will start a program. Then there will be an election. Politics will change. Politicians will put somebody else in charge of NASA. The budget will be cut then plussed back up. Different personalities and lobbyists will pressure Congress to emphasize different missions and directions. NASA has no long-term planning at all. In fact, we as a nation have no long-term plan. With the rate our technology was growing in the 1960s and '70s we should have colonized the Moon by now and put humans on Mars. But we have had no plan to do anything.

What a waste of effort, time, and elbow grease. I can't imagine doing a weekend build without an end goal in mind. NASA has a serious lack of focus and attention deficit disorder. Five Rocket City Rednecks accomplish some pretty awesome things in one weekend with minimal budget usually because we pick a goal, make a plan, and *STAY WITH IT!*

NASA and the nation's space community need to learn how to pick a goal, make a plan, and stay with it.

★ ★ ★

So, the Obama administration decided that it would be a brilliant idea to kill the space plan that Bush had, simply due to politics. To throw out the few billion dollars and five or more years' worth of work, and to start over due to strong lobby pressure from Lockheed Martin, Boeing, and SpaceX, and create a new commercial launch system approach and a Space Launch System. This, after we'd already had a successful, full-up launch of the Ares-1X prototype rocket.

Never mind, we'll start all over again. Well, OK, that's great, all well and good. A fresh start. Again. For the umpteenth time! But what are the components of this *new* space launch system? This new idea could've been great and based on American derived parts. It could've been based on the Space Shuttle Main Engine, for one thing. It could've been based on the RS-68 used on the Delta IV. It could've been on the J-2 engine from the Apollo era and Constellation program. It could've been a new American space engine. It could've been many other engines.

Originally, through administration directive, the replacement plan was to use the Atlas-5 engine. The Atlas-5 is an American rocket. That's a great idea. It's a wonderful rocket. But what we don't seem to broadcast to the general public is that the Atlas-5 engine is based on a Russian engine—we have a license from them to use it. Once again we would depend on the Russians for our space launch capability.

That's bizarre. That is insane. And it is downright un-American. We should be ashamed! Every time we will want to go into space in the future we will have to pay the Russians a license fee? That is just crap. Fortunately, at least at the time of writing this book, it looks like the NASA engineers, Shuttle and Constellation program contractors, and common sense may win out and, instead of the Atlas-5 based system, the space launch system will be more Shuttle derived. More on this later.

I mentioned the Constellation program *en passant*. So what was it?

Not too deep into the Bush administration, GW proposed that we would go back to the Moon by 2020 and then on to Mars. The crew compartment was known as the Orion capsule, and the rocket was the aforementioned Ares. Together they comprised the Constellation program.

But GW only slightly ramped up NASA's budget. And the administration still expected NASA to continue, with the Space Station and Shuttle program also funded out of that same budget. The Constellation program was a brilliant idea. In fact it was a combination of an idea that Wernher Von Braun had had with modern Space Shuttle derived components. It also applied some Apollo-era style space capsules and landers. Therefore, the Constellation program concept was safer than it had been previously, based on heritage ideas and components and tests and flown vehicles. And it was actually a goal that was achievable even though the budgets were really slim. But when President Obama was elected, the first thing that happened, due to politics, was that the program was killed so that we could start something different.

Now the Constellation program, while it got people enthusiastic and full of plans, it flatly didn't have the budget it needed. A lot of people think that NASA's budget is huge and if we are spending too much money in the first place, it's because NASA spends so much money.

That's bogus. And it shows how little Americans understand about the economics of our nation. If you draw a pie chart of America's budget and you put wedges representing the pieces of pie for all the different programs that we spend money on, NASA is barely even visible. All you can see is that the defense budget is a huge chunk, but the biggest chunk is our entitlement programs for welfare and such. But NASA is a single tiny line so small you can barely see it with the human eye on a pie chart that'll fit on a regular piece of notebook paper. NASA's trying hard with the budget it has, believe me. In fact its average is somewhere between ~$13–$15 billion a year ever since it started. The Apollo era program budget was on the order of ~$13–$15 billion a year. And that was in the 1960s. During the Space Shuttle era, it was between ~$13–15 billion a year. And that was in the '70s and '80s. The International Space Station and everything else in between was ~$13–$15 billion a year. And that was in the '90s! The Constellation program through the early 2000s had to fit into a NASA budget of ~$13–$15 billion a year while *still* keeping the Shuttle and Space Station going with that same money. And it's still ~$13–$15 billion a year in the year 2012.

Think about that. It's still ~$13–$15 billion a year in 2012. Now, if you have ever heard anything about economics, if you ever bought anything in your life and then bought that same thing a few years later, you understand what I'm getting at. You don't have to be an economic genius to see the problem here. NASA's budget, number-wise, dollar-wise, has stayed the exact same number pretty much forever. But the dollar has not stayed at the same value. $15 billion a year in 1965, depending on which inflation model you use today, would be somewhere on the order of $200–$250 billion. Some models on the lower end suggest about a factor of ten so that's $150 billion. Even using that lowest scale of $150 billion a year, that means that NASA's budget has been cut by at least a factor of ten since the Apollo program. And Congress and the Bush administration expected the NASA engineers and scientists to do a *new* space program to the Moon, maintain the Space Shuttle, *and* complete the construction and operation of the International Space Station, for ten times smaller a budget than the single Mercury/Gemini/Apollo era programs had in the 1960s. And remember that's the budget model using the *lowest* inflated dollar values. To use some of the other models means NASA's budget has really been cut by a whole lot more than ten times. Whatever the value is, it is clear that NASA's budget has been reduced dramatically such that they have no choice but to become a starving, failing organization.

But you know what? The engineers and scientists and technicians at NASA, to their credit, had figured out a way, a plan to get us back to the Moon by 2018, keep the Space Shuttle flying to 2012, and maintain the operations of the International Space Station. On that skimpy $15 billion a year budget!

Oh, there were some problems, and a little bit of cost overruns here and there, but nothing dramatic on a program scale. And there might've been some lags where we didn't have the Space Shuttle to get us to the Station for a year or two. But now we do not have a way to the Space Station for many years to come unless some significant investment is put back into the space program. Once again, I am keeping in mind that pie chart of America's budget when I say "significant investment into the space program." I don't mean significant like the big pieces such as the entitlement budget. If you just increase NASA's budget tenfold to $150 billion a year it would still only be a line that looks like it was drawn by crayon as opposed to a thin lead pencil. If you don't believe me, try that for yourself. Go and draw out America's budget in a pie chart and put a wedge on the scale for the $15 billion and then put another wedge on the scale to $150 billion. What you will find is that the difference is barely noticeable compared to the other big pieces of pie.

2012 U.S. Budget [4]		
Submitted:		
On 02/14/11	by Barack Obama	to 112th Congress
Passed by Houses of Congress:		
November 18, 2011 (Pub.L. 112-55)	December 23, 2011 (Pub.L. 112-74 and Pub.L. 112-77)	
Total revenue $2.627 trillion (requested)		
Total expenditures $3.729 trillion (requested)		
Deficit $1.101 trillion (requested)		

4 http://en.wikipedia.org/wiki/2012_United_States_federal_budget

NASA Budget Comparisons[5]		
2012 NASA Budget	7.8 billion dollars	
Percent Federal budget	0.48%	
Percent 2012 Federal deficit (calculated)	0.71%	
Percent 2012 Federal revenue (calculated)	0.30%	
Percent 2012 Federal expenditures (calculated)	0.21%	
Largest ever NASA percentage of U.S. Budget	4.41%	Year: 1966
Largest NASA percentage of U.S. Budget during Shuttle operational time frame	1.05%	Year: 1991
Largest NASA Percentage U.S. Budget between Apollo/Soyuz mission (1975) and STS-1 (1981)	0.99%	Year: 1976
NASA Percentages of U.S. Budget during the years between Challenger disaster (1986) and Return To Flight (1988)	0.75%	Year: 1986
	0.76%	Year: 1987
	0.85%	Year: 1988

[5] Sources for budget numbers: Teitel, Amy (2011-12-02).
"A Mixed Bag for NASA's 2012 Budget." DiscoveryNews.
http://news.discovery.com/space/nasa-2012-budget-ups-downs-111202.html
http://en.wikipedia.org/wiki/Budget_of_NASA#cite_note-8

★ ★ ★

This is where the education in this country has failed us. We haven't taught people how to understand America's budget. That's how we can get trillions of dollars in debt and have not a dadgum thing to show for it. That's how we can think that service industries and virtual industries are products that will make America strong. That's how we can get into a situation where we believe that it's okay to buy everything that we need from China. That's how we can get into a situation where we don't have enough students becoming engineers and scientists in this country to build a danged phone much less a space program.

When I talked the guys into starting the TV show on National Geographic Channel called the *Rocket City Rednecks*, it wasn't just so we could be on TV. Our goals were not all one hundred percent altruistic, of course. But at least fifty to seventy-five percent of our goal was to get kids out from in front of video game consoles and get them out into the garage and building things. We've got to get our kids excited about doing science and engineering and building gadgets and doing experiments and making this country great, building the things that make us great. Like rockets that will take us back to the Moon. That's exciting! That's inspiring!

So what we need is a new American space program. One that we can stick with. We need a new American space program that is based on the goals and requirements for achieving those goals, not on whatever political opinions are in favor and the changing of administrations. We need a new American space program that has longevity and can't be monkeyed with by new politicians, new presidents, congresses, the UN, China, Russia, or anybody else, for that matter. We need to set the plan in stone and stay with it. We need to make a decision as a nation that we're going to be number one in space and stay number one in space—and then *do it*. We need to put it into the law for the program be NASA or not, commercial or not, that this is the budget for this many years until the goal is achieved and leave it at that and say hands off.

CHAPTER 4:
HOW WE MADE THAT
"ONE GIANT LEAP"

The American space program used to be full of nothing but doers. Then somehow we lost all that and became a whole bunch of PowerPoint Rangers. Go find four people at NASA who have actually built a rocket part with their bare hands and then set it off and you'll find four really old dudes that should have retired long ago. After all, we've "outsourced" our space program to the Russians. I'll get on that soapbox in another chapter. Of course, I'm exaggerating about finding four actual NASA employees that have built rocket parts to make the point. There is probably a handful more. Also, it ain't NASA's fault any more than it is yours or mine because the culture of risk aversion that has overrun the country has infected everybody. Well, I take that back, I guess it *IS* our fault for letting our country become a nation full of folks sitting their big butts on the couch and playing video games and watching Oprah instead of getting out in the garage and building a go-cart or working on a motorcycle or building amateur radios or inventing the next microcomputer.

But the NASA of my daddy's time . . . well, it was different. We were nationally driven to beat the Russians and to live up to a great president's dream. America was a nation full of doers back then.

The most likely place to see noncontractor labor having built something at a NASA facility these days would be to visit the U.S. Space and Rocket Center during the Great Moon Buggy Race. This is a college and high school competition for kids to build man-powered buggies inspired by the lunar rover used in the Apollo missions to the Moon. These buggies then compete on a very harsh obstacle course that simulates the rigors the astronauts saw on the lunar surface while in their rover. The college and high school kids come up with and build some great stuff. Thank God they got off the couch and put down the Xbox for a little while.

Actually, I guess as the kids aren't employed by NASA they don't count as NASA doers . . . yet. We can still change all that. We need some big emphasis to put these bright kids to work. NASA needs to start hiring again with a NATIONAL goal driven by the moral imperative of the U.S. being first in space. But they aren't really hiring right now. No, not at all. Go to USJobs.com and look. You might find a handful of positions. Most of them are for managers and contract specialists. There is an occasional scientist/engineer job. Why? Because NASA has no budget and no goal to put our bright kids to work on. Somebody has to keep these kids invigorated, enthusiastic, and thinking.

So, I decided one weekend that the Rednecks were going to do the Great Moon Buggy Race. But we were going to do it differently. We were doing it in spacesuits because that is how astronauts would have to do it. I wanted the kids to realize this through our demonstration of it. Being in a spacesuit changes how you would design such a thing. This was a good manned-space hardware engineering demonstration for the kids as well as us.

I'll talk more about this later, but the main point is that we wanted to build this thing in a very short timeframe, with few dollars, and potentially be able to use it on the moon for real. We did use inflated tires because we didn't have time to make solid ones, but hey, it was just a prototype. A prototype! This is why I bring this up here. NASA should be building prototypes. Government labs in all branches of the government should be building prototypes.

The first lunar rover wasn't a big program or planned out thing in detail with hundreds of PowerPoint Slides, meeting after meeting, and months of government acquisition steps to be taken. Heck, PowerPoint wouldn't be invented for about thirty years. No, the first lunar rover was made by a couple of guys using spare parts in the back of a machine shop in Huntsville, Alabama. When the first prototype was complete they had yet to put chairs or seats in it. What did the engineers do? Did they have a meeting, draw something in 3D on their computers, or fill out a bunch of forms, or write up a contract for some contractors to work on it?

We built a man-powered Moon Buggy in two days that competed with some of the best in the competition. We raced ours in spacesuits to demonstrate to the youngsters that if it were a real lunar rover there are a lot more things to consider when building it.

No they did not. They walked down to the lunchroom and sawed the legs off of a couple lunchroom chairs and welded those jokers onto the prototype rover. So, to pay tribute to those doers we used lunchroom seats on our Redneck Moon Buggy also. In the end, it was only the wheels that caused us a problem later, but you can watch that episode for those details. With another weekend or two and a couple more hours of testing we could have likely built a Moon buggy that we, while wearing bulky spacesuits, could use to beat the kids in their biking shorts.

Now, I could name specific names at NASA who laugh at this and wouldn't hire us Rednecks to save their lives because we'd give NASA a bad name simply because we don't do it the NASA way—make PowerPoint slides and have meetings ad nauseam. Oh, but we all know how to do that. Making slideshows is easy, but we also like to do things. Do things! NASA needs to get back to doing things. I do have to say here that the U.S. Space and Rocket Center folks have been more than happy to work with us. They have a shoestring budget to keep one of America's most important museums alive and they get what we are doing and we get what they are doing with the show. But they don't really count as the modern NASA.

We made it all the way to the last obstacle when both mountain bike wheels on the driver's side collapsed! While pushing the buggy across the finish line we exclaimed that we'd never use those kind of wheels again.

These modern day NASA managers would throw the original lunar rover prototypers under the bus and run over them with it a few times rather than praise their innovative efforts. That is how much NASA has changed. Oh, I'm not bashing NASA here as a whole. I am actually a big supporter and proponent of all things NASA. People at NASA still want to build stuff and do stuff, but the culture is completely politicized and so risk averse as to prevent and even stifle hands-on creativity that got us to the Moon and made NASA famous in the first place.

It is the glimmer of hope in things like the Great Moon Buggy Race that shows me that the new generation out there still doesn't mind getting their hands dirty just like their spacefaring grandparents did— or is that *great* grandparents now.

Whatever the case may be we need to relearn from history how we made it to the Moon the first six times. We need to learn how to build things again. We need to quit having so many dang meetings just to have meetings and mark off steps in an acquisition process that is killing creativity and productivity and our great country. We *NEED* to hire a bunch of Moon Buggy builders and let them build some dang Moon buggies!

We also need to allow for some failures and some partial successes. This is how we learn. Our big Double Barrel Rocket, as it was called, failed spectacularly due to some missing parts in the solid motor casings, we think. We learned a lot from that build. We learned that some of our structure was spot on. Some of it not so good and some of it was over-engineered like the Roman roads. The biggest thing we learned was that it is okay to fail and that when it comes to experiments no flight test is a failure if you learn something. This lesson needs to be addressed in the current NASA culture. Politics puts unneeded stress on an already stressful experiment. The space vehicle can't fail or the funding will go away. We have to realize that sometimes we learn from such "failed" flights. Before the first man was in a Saturn V there were over twenty unmanned flight tests of all sorts. Some rockets failed; some didn't. But, all the flight tests were successful because we learned how to step forward and get to a better design for the next test.

These are the types of lessons that the coming generation needs to learn from things like our show and from the Great Moon Buggy Race. Sometimes exploding flight tests are spectacularly successful.

★ ★ ★

What makes sense about our past space plans? Clearly, from the beginning of our space program in the early '60s, we had the moral imperative and drive to be first in space before the Soviet Union beat us there. There was a serious cold war taking place and there was the fear that whoever owned space would be the winners of the Cold War. History shows us that that was not too far from the mark. It was our eyes in space and missile technologies and communications satellite technologies that gave us the advantage over the Soviet Union. Of course the economics of the Strategic Defense Initiative is probably one of the main things that toppled the Soviet Union. But that wouldn't have been as effective had there not been such a drive for superiority in space.

The original plan to put man on the Moon as soon as possible really invigorated America and America's future and children. The first approach was to find some very smart guys who could build rockets. Well, we found those guys at the end of World War II. The German rocket scientists that we had found were better than any others in the world at the time. Those Germans were brought to White

Sands, New Mexico for a few years and then moved to Huntsville, Alabama on the Redstone Arsenal. They weren't much loved back then because many of them were instrumental in killing a lot of people using their missiles that rained on London. And there was an attitude in America at the time that there just weren't any good Nazis.

Whether the German rocket scientists' motives were evil or pure, forced or voluntary, is immaterial. They were the best rocket scientists on the planet and we needed rocket science. Rockets are very complicated devices. They have many explosive components, instrumentation components, and moving parts. In some cases they have hundreds of thousands to *millions* of moving parts.

Clearly, a handful of German scientists can't build a complete rocket system with so many components. They can't all by themselves make manned spacecraft orbit the Earth, or eventually land on the Moon and safely return. Throughout that entire process are tens of millions of parts and components to be manufactured and integrated together. It took hundreds of thousands (maybe even millions) of people to do that—Americans. And the rockets that were constructed and tested were tested right here in the Rocket City—Huntsville, Alabama—by a community that before the space race were a bunch of sharecropping farmers—also known as rednecks. These rednecks put down their plowshares for a little while to help the Germans construct this American space program. Hopefully, history will be appropriately recorded and will show that the original American space program was conducted by a handful of German rocket scientists and a town full of Rocket City rednecks.

The first American in space, Alan Shepard, was launched on a Mercury-Redstone vehicle on a suborbital flight. A Redstone rocket was an Army missile. The rocket used ethyl alcohol (ethanol) propellant —drinking alcohol—that had about a twenty-five to seventy-five percent water/ethanol mixture so that, when mixed with liquid oxygen, an oxidizer, it would burn very well as a rocket fuel. Having the Redstone running off pure grain alcohol, with the high explosive nature of the mixture, was nothing that was uncommon for the locals in the community. My dad tells stories about my distant relatives who might have been, wink, wink, employed by the moonshine trade. I'm stressing the word "might" here.

The second American in space, Gus Grissom, also flew on a

Mercury-Redstone rocket. This was again a suborbital flight and only lasted about fifteen minutes or so. The little Redstone rocket just couldn't push the space capsule into orbit.

The first orbital American flight was made by John Glenn. He rode atop a USAF Atlas LV-3B. The Atlas was big enough to push the capsule into orbit for the first time. It was a "one-and-a-half-stage" rocket, unlike the Mercury-Redstone, which was single-stage. Also unlike the Redstone, it used a fuel called RP-1, a refined petroleum fuel similar to jet fuel, with LOX for the oxidizer. This produced nearly four times more thrust and higher specific impulse, too (specific impulse is thrust divided by the rate of fuel burned, and to a rocket scientist is sort of like telling the miles per gallon a car can get).

America was on its way to the Moon. And we finally knew how to put men into space and keep them there, at least orbiting the Earth. The next step was to learn how to walk in space, and to meet up with other folks in space, and carry larger payloads and equipment and more people on a single vessel. This led us to the Gemini program. The Gemini program used a modified version of the Air Force's Titan II rocket system to launch a two-man capsule into space. A two-man capsule would be in orbit and mate up with an unmanned vehicle called the Agena module. This part of the American space program was to learn how to dock two vehicles. The reason for that was we hadn't quite figured out exactly how we were going to get all the equipment up to the Moon. There were two trains of thought at the time.

Two vehicles would launch the equipment, and they would meet up in space and go from there to the Moon. This is what the Constellation program was gonna do before the program was canceled by the Obama administration.

The other thought was to build one big rocket vehicle and launch everything into orbit all at once, and all of the equipment could then go on to the Moon once they got to Earth orbit. It sounded simpler at the time.

It turned out that the docking was required either way. Once the Apollo service module and the lunar excursion module (LEM or later just LM, for lunar module) were in space and deployed from the Saturn V launch vehicle, we actually had to turn the lunar excursion module around and dock it with the nose of the service module before we went to the Moon. And once at the Moon the lunar excursion module would land on the Moon. And eventually when the astronauts were ready to go home they had to fly the lunar excursion module back up to the service module and dock with it. Hence, the Gemini program was well thought through and we found out a lot of different things about operating in space from it.

There was a method to the madness, see. Project Mercury was designed simply to get astronauts into space and demonstrate that it could be done, and done safely, and get them home in one piece. Project Gemini was more ambitious. It had to look ahead to Project Apollo and determine what equipment and techniques would be needed to accomplish that project, then develop, validate, and practice them. That included extra-vehicular activities, aka EVAs or spacewalks; six-degree-of-freedom maneuvering (up/down, right/left, forward/reverse, pitch, yaw and roll), rendezvous and docking, and simple things like extended duration flights, coordinating the activities of more than one guy in a confined space, and mission-essential details like that. Project Gemini was one of those steps that, at first glance, you'd think you could skip—until you realized that it contained all of the necessary development and practice to do the *next* step. That is the kind of practice that more lunar missions would be, to later get to Mars or an asteroid.

Once we figured out how to work in space and dock in space it was time to go to the Moon. That's when we started building the big Saturn V rockets here in North Alabama at the rocket center. My dad worked in detail on those things, and he can tell you all about the glory days of Americans going to the Moon. And he will if you ever give him more than two seconds to catch his breath and spin up a story.

The Saturn V rocket was a big mother. The first of three stages had five F-1 engines, the biggest rocket engines ever built by Americans *and* fired. It had thirty-four mega-newtons of thrust, which equates to over 7.6 million pounds of thrust! It used RP-1, a kerosene-derived fuel similar to jet fuel, with LOX, just like the Atlas had that carried John Glenn into Earth orbit. The second stage had five J-2 engines, and the third had a single J-2 engine. These latter two stages used liquid hydrogen and liquid oxygen, like the Shuttle Main Engines, as fuel and oxidizer. If they were fired consecutively (including the firing that ejected the Command Module from lunar orbit), it would take less than seventeen minutes to empty the fuel. The entire stack was 363 feet (110.7 m) tall but only 33 feet (10.1m) in diameter. It weighed 6.7 million lb (~3 million kg) and was capable of carrying 262,000 lb (119,000 kg) to LEO, or 100,000 lb (45,000 kg) through a trans-lunar injection trajectory.

Once we got the Saturn V rocket working we launched and successfully landed twelve men on the Moon, two men each, six different times, while each time there was another man in the service module orbiting the Moon. Though we managed to do a lot and bring home some great samples and learn a lot about the Moon, we didn't see much of it, because a man in a spacesuit can't walk too far from the landing site. And even those missions where we had a lunar rover, we were limited to not traveling farther than about 4.7 miles away from the lander. This was because if there was a rover failure then the men would have to walk back and there was only enough life support in the suit for six hours. That had to be enough time for them to walk back from wherever they'd gotten to. The lunar rover was helpful but very limited in the range it allowed the astronauts to travel.

Now that's not to low-rate the lunar rover, mind you. The lunar rover was invented right here in the Rocket City. The first one actually used lunch room seats on the prototype. That prototype was pretty much made out of spare parts and junk that was lying around because all the good components were being put on the rocket. But once the prototype was constructed and demonstrated, it was clear that the astronauts needed it and would get it, so they put out a request for bids and awarded a contract to build 'em. The rover did manage to allow astronauts to cover much more territory. But think about a circle that has a radius of 4.7 miles (~7.6 km).

So in six different landings, at best we got to see approximately sixty-nine square miles (182 square km) of the Moon, total. That's all humans have seen and experienced themselves. But the total surface area of the Moon is 1.37×10^7 square miles (3.79×10^7 km^2). That means that the astronauts saw a total of about 0.00048 percent of the surface area on the Moon, and that's a generous estimate. In other words, a drop in the ocean. We really haven't seen any of the Moon. We know very little about it. And for politicians to say that "we don't need to go back to the Moon," and that "we've been there and done that" shows how little they know about our space exploration program.

And how bad our public education regarding space exploration is, for the public to accept that statement. Such statements are nonsense. We know very little about the Moon and need to go back.

This would be a good time to take a look at how we went to the Moon the first time. That means we need to look at the rocket that got us there, and the rover that drove us around.

The Saturn V launch vehicle was one of the largest man-rated rockets ever built, and the largest ever built by the United States. It stood 363 feet (111 m), or thirty-six stories, tall, and had a base diameter of 33 feet (10 m). When fully fueled for launch, it had a mass of 6.2 million pounds (2.8 million kg). How heavy is that? If we had a gigantic scale, and put the Saturn V on one side, it would take four hundred elephants on the other side to balance the scale. The full Saturn V rocket with all stages could launch 130 tons (118,000 kg, or ten school buses) into low Earth orbit, but only 50 tons (43,500 kg, or four school buses) to the Moon. The difference is due to needing the extra thrust to escape Earth's gravity well.

It was developed right here in the Rocket City at the Marshall Space Flight Center, along with its smaller siblings, the Saturn I and IB, which were designed for injection into low Earth orbit. The first Saturn Vs were test rockets, and were flown unmanned. The first manned Saturn V launch was Apollo 8, which orbited the Moon but didn't land. The first landed mission was, of course, Apollo 11.

The Saturn V had three stages. The first stage, called the S-IC, generated 7.6 million pounds (34.5 million newtons) of thrust. It used five F-1 engines running on RP-1 (modified jet fuel) and LOX. The center engine was fixed, but the outer four could gimbal to help control direction. The amount of fuel it used would power a thirty-mpg (12.6 km/L) car for 800 laps around the Earth. It generated more power than eighty-five Hoover Dams, maxed out and running simultaneously. And that was just the first stage!

The S-IC burned for a full 150 seconds. That's two and a half minutes. And then its fuel was exhausted. By that time, it was about forty-two miles (67 km) up, and fifty-eight miles (93 km) downrange of the Kennedy Space Center, moving at 7,500 feet per second (2,300 m/s).

Then the S-IC was jettisoned along with the aerodynamic fairing connecting stages, venting excess fuel as it fell toward the ocean, and the second stage, the S-II, opened up. Like the first stage, it had five engines, with the outer four being gimballed, but these engines were J-2 engines, powered with LH2 and LOX. It was the largest cryogenic rocket ever built until the Space Shuttle. It had a thrust of 1.1 million pounds (5.1 million newtons). It burned for six minutes, at which time the craft had reached an altitude of 109 miles (175 km) and was traveling at 15,647 mph (25,182 km/hr or 7 km/s), close—but not quite—to escape velocity.

The S-II was then jettisoned. It, too, vented fuel en route to the ocean far below, and the S-IVB ignited. It used a single J-2 engine, again running on LH2 and LOX. It had a thrust of 225,000 pounds (1 million newtons).

This third stage had a multiple purpose. It burned for about two and a half minutes until first cutoff at Mission Elapsed Time (MET— duration from launch) of eleven minutes, forty seconds. By this time the craft was 1,640 miles (2,640 km) downrange of the Kennedy Space Center, at 118.8 miles (191.2 km) altitude, moving at a speed of 17,432 mph (28,066 km/hr)—still not quite escape velocity. But they didn't want that yet. That would come a bit later, when they'd reached the correct part of the orbit.

This was a very low orbit, and atmospheric drag would affect it quickly, but it was considered a "parking orbit," a temporary orbit only needing time to prep for the trans-lunar injection burn, which would also be performed by the S-II. Also, the S-IVB maintained a low-level thrust in order to prevent cavitation (bubbling) in the fuel lines, keep the fuel tanks settled, and vent excess hydrogen as it boiled off from the cryogenic liquid. This more than compensated for the drag.

When they reached the right part of the orbit (which varied from mission to mission, but for Apollo 11 was at MET 2 hours, 44 minutes), the S-IVB opened up again for the Trans-lunar Injection (TLI) burn. This lasted about six minutes, at the end of which time the craft had reached Earth's escape velocity of 25,053 mph (40,320 km/hr or 11.2 km/s).

At MET 3 hours, 24 minutes, the Command Module separated from the S-IVB, turned 180°, and docked with the Lunar Excursion

Module stowed in the upper part of the S-IVB. The Command Module extracted the LEM, and the S-IVB had only one more task. It vented excess fuel and ignited its auxiliary propulsion system to take it out of the Command Module's trajectory, to avoid collision hazard either in lunar orbit or upon return to Earth. On the earlier Apollo missions, it was directed at the opposite edge of the Moon and into a slingshot trajectory into space, to orbit the Sun. Later missions directed it to crash into the lunar surface. This was not merely waste; it had a purpose—earlier missions had left seismometers on the Moon, which could detect the impacts, thus enabling scientists to get data on the interior of the Moon.[6]

There was, of course, lots and lots of testing, especially in the early days. After all, we were on a short time schedule and we were going to be putting real men on top of this thing. Testing did not always proceed without incident, although it wasn't always quite the incident expected. In 1964 the S1C static test stand was completed at Marshall Space Flight Center. This test stand was for live firing of what was, essentially, the entire Saturn V first stage—all five engines, over 7.5 million pounds of thrust. The normalized sound level of the first stage firing is about 220 decibels. By contrast, this is 5 db louder than the loudest thunderclap, and roughly equivalent to the sound of the larger of the two WWII atomic bomb blasts. The rumbles regularly produced earthquakelike shaking, and could be heard as far away as 100 miles (160 km). That last bit is the official line, but the guys who were working there, back in the day, including my dad, told me there was a little more to it than that.

In general, whenever such engines were fired, whether it was for testing or launch, there is always a careful check of weather conditions. Storms in particular, with heavy lightning, are feared, as the combination of conductive materials and high explosives with lightning is not a good one. High winds are also not favored for reasons of blowing debris into sensitive parts, or during launch, blowing the spacecraft off course.

On one of the early tests, however, the weather conditions were

---

[6] http://www.nasa.gov/audience/foreducators/rocketry/home/what-was-the-saturn-v-58.html

http://en.wikipedia.org/wiki/Saturn_V

perfect except for a low, solid overcast. A stratus cloud layer, covering the entire region, had moved in, but was causing nothing except for lowered lighting levels, and so the decision was made to proceed with the test. The team went down through the countdown checklist, and the first stage was fired, quite successfully according to all accounts.

And then the reports started coming in: people in Birmingham scared to death, the ground moving, buildings shaking, windows blown out in certain parts of Birmingham (which, by the way, is about 100 miles away), and finally, the seismologists of the New Madrid Seismic Zone contacting NASA to find out if anyone had felt the earthquake that seemed to have its epicenter in northern Alabama, awfully close to Huntsville.

Turns out that the cloud deck acted kind of like a wall, and it reflected the atmospheric vibrations from the engine test. It set up standing waves that radiated in all directions, with a peak happening to occur . . . yep, you guessed it, in Birmingham.

After that, a new constraint was added to the weather conditions acceptable for testing and launch. No solid overcast testing.

The lunar rover, aka the Moon buggy, became an essential piece of equipment on the later Apollo missions. Now, working in a pressure suit such as a spacesuit is very difficult. This is partly due to the thickness of the suit, but it is also because of its very nature. It is like working inside a balloon. You might have seen pictures of the tools astronauts use, and noticed their thicker than usual handles, even on a standard object like a hammer. This is because it is next to impossible to be strong enough to, say, make a fist wearing a pressurized space-suit gauntlet. You're pushing against the very pressure of the suit. An astronaut on EVA who has to do a lot of physical activity wears out very fast. So the rover made it possible for astronauts to go farther, faster, with less exertion. They were utilized on Apollo 15, 16, and 17. A fourth was also built, but used for spare parts after further Apollo missions were cut. There were also a number of test models built for various uses. In three lunar missions, the vehicles racked up a total of over fifty-four miles (90 km) on the Moon. The longest single traverse occurred on Apollo 17, when the astronauts took their rover over twelve miles (20 km) in a winding, roundabout path. But they never got more than 4.7 miles (7.6 km) from the lunar module.

The rover is actually a very complex vehicle, and its development was coordinated out of Marshall here in Huntsville. Since typical combustion engines require oxygen—an atmosphere—to work, it was electrically powered, with each of four wheels having its own electric drive, identical 0.25 hp (188W), DC series-wound motors, capable of 10,000 rpm each. Each was attached to its wheel with an 80:1 harmonic drive. In addition there were two more motors, used for front and rear steering. These were DC series-wound as well, 0.1 hp (75W). All six, plus a 36V outlet for the TV camera, were powered with two 36V, silver-zinc-potassium hydroxide batteries, which together supplied 242 Ah. That made for a range on the rover of fifty-seven miles (92 km).

The frame was constructed of aluminum alloy 2219 tubing. This was welded into a three-part hinged chassis, to enable it to be folded and stowed in a bay of the Lunar Module. When unfolded, it measured 10.2 feet (3.1m) long, 3.74 feet (1.14m) high, with a wheelbase of 7.5 feet (2.3m), and had a steering radius of 10.2 feet (3.1 m) if the independent drives for front and back steering were used. In other words, it had the capability of turning inside its own length. When fully loaded (the rover could hold 1,080 lb (490 kg) of payload, including two astronauts), it had a ground clearance of fourteen inches (36cm). Its own weight (on Earth) was only 463 lb (210 kg). On the moon, it weighed about 77.2 lb.

It had a horizontal double wishbone suspension, upper and lower torsion bars, and a damper unit between the upper suspension and the chassis.

The wheel/tire assembly was 32.2 inches (81.8 cm) in diameter and 9 inches (23 cm) thick. The hub was made of spun aluminum, and the tire was a strange contraption. It had an inner frame, only 25.5 inches (64.8 cm) in diameter, called a "bump stop." This protected the wheel hub in the event of lunar "potholes," usually known as craters! The outer part of the tire consisted of what was essentially a zinc-coated, woven steel belt (the steel strands were only 0.033 inches 0.083 cm thick!) that looked sort of like sophisticated screen wire. On this, the builders placed chevrons of titanium in a pattern that covered only half the surface area of the tire—this was the "tread." Dust guards, like fenders, were positioned above each wheel, as the earlier Apollo astronauts had already noted the tendency of the fine lunar dust to work its way into everything and kick up quite easily.

Unfortunately on both Apollo 16 and 17, one of the rear fenders broke upon being struck, the former by an astronaut, the latter by an astronaut's hammer. Evidently no attempt at repair was made on 16, and the crew, cargo, and control panels became absolutely covered in lunar dust. But on 17, the astronauts tried to tape it back together—using, no doubt, the universal remedy for anything that needs sticking together, the ubiquitous "gray tape" aka duct tape. (The more general term "gray tape" is used whenever possible to avoid the use, and hence possible interpretation of endorsement, of a particular brand.) However, by that time the fender was so dusty that the tape wouldn't adhere, and the broken part fell off while the rover was returning to the LM. Astronauts being astronauts, they got creative. Some more gray tape, a couple spare EVA maps, and a couple of clamps later, they had a makeshift fender that worked fairly decently for the rest of the mission, at which time they removed the clamps and maps and brought them back inside the LM for return to Earth.

The luxurious passenger space of the rover seated two, in folding seats made of aluminum tubing and nylon webbing, not unlike a folding picnic chair, but sturdier. A "center console" armrest was between the seats, and each seat was outfitted with adjustable foot rests and seat belts that velcroed closed. The floorboards were aluminum panels. Steering was provided by a T-shaped hand controller between the seats, which controlled each of the four wheel motors, both steering motors, standard brakes, parking brake, and gear (forward/reverse). The thing could be driven with one hand. The equivalent to a dash console displayed not only an odometer and speedometer, but gauges for direction, pitch (up/down hill angles), power, and temperature.

Primary navigation was based on continually monitoring the odometer and direction measurements (the latter provided by a gyroscopic system), and inputting the data into the onboard computer, which then told the astronauts how far, and in what direction, the LM was. Secondary navigation used a manual sun-shadow device.

Mounted on the "hood" of the rover was a color television camera that could be remote controlled from Houston. It could be panned, tilted, and zoomed, which meant the astronauts could concentrate on exploration without having to constantly mess with the camera—the guys on the ground could control that. As it was, much of the equipment on board the rover were cameras of various sorts anyway, still and video, to record their surroundings in detail.

Also on the "hood" was the large mesh dish high-gain antenna that communicated to the Command Module in orbit overhead, which then relayed the signal back to Earth, and vice versa.

To deploy the rover, the quad 1 bay of the LM, where it was stowed in a folded, hanging configuration, was opened. One astronaut climbed the LM's egress ladder and released the rover, and the other astronaut lowered it carefully with a system of pulleys and reels. The unfolding was automatic: the rear wheels unfolded and locked, then eased to the ground. Then the front of the rover would be unfolded, locking into place, before the front wheels also unfolded and locked. The astronauts completed easing the rover onto the lunar surface, and then removed the various cables, pins, and such like that had held everything in place. All that was left was to unfold the seats, raise the foot rests, power it up, and back away from the LM.

The final cost of this amazing car that was designed to drive on other worlds? $38 million.

". . . the lunar rover proved to be the reliable, safe and flexible lunar exploration vehicle we expected it to be. Without it, the major scientific discoveries of Apollo 15, 16, and 17 would not have been possible; and our current understanding of lunar evolution would not have been possible."

—Dr. Harrison "Jack" Schmitt
Geologist, Astronaut, Apollo 17[7]

Okay, fine. So what's the big deal? We just do it again, right? Build another Saturn V, another Apollo capsule, another LM, another rover, and go back. Simple.

Except we can't.

The rub here is that when the Apollo program ended, it completely disbanded. All of the facilities for making all of the parts of a Saturn V were taken offline and either torn apart or repurposed. The people

---

[7] http://nssdc.gsfc.nasa.gov/planetary/lunar/apollo_lrv.html

who made all that stuff moved on to other jobs, eventually retiring. They took their skill sets, their knowledge, and even the plans and paperwork with them when they retired. Many are no longer with us, boxes in attics, by well-meaning relatives, all unknowing of those papers' importance. We literally no longer have the capability to duplicate the Apollo program.

We can't just reproduce Apollo. We can go to the Moon again with more modern technology, but we'll get to that in a bit. First, we should consider what other countries are doing and planning to do when it comes to space.

CHAPTER 5:
WHAT IS THE REST OF THE WORLD PLANNING?

I recall when I wrote my first science fiction novel that there was a scene I did on the International Space Station. Something was broken on an experiment and the Americans and the Japanese couldn't figure it out. I had a Russian character fix it with a stick and some duct tape or something along those lines. I can't count how many times on Rocket City Rednecks a stick and some duct tape did wonders for solving a problem. This reminds me of the Russian Mir space station where things were often badly falling apart, but the Russians used some spit, chewing gum, and baling wire to keep it afloat. I like the spirit of the Russian space program because they have little money but a lot of heart. Even if something breaks they figure out a way to keep it moving in a very frontier sort of way.

This can-do spirit is what the Americans had during the Apollo era. Had we not had it then Apollo 13 would likely not have returned home safely. But what was driving the American program was fear of being overtaken by the Russians. So it is only fit to spend a good deal of time talking about the Russians in this chapter. There is something very appealing to someone using stuff that is lying around and making something useful from it. Something very MacGyver about it.

This reminds me of the Spy Satellite episode we did using a latex weather balloon. We needed a communications antenna and hadn't thought ahead enough to buy one. Sometimes on the set there is just no time to think ahead. We've been bitten by that issue more than once.

So, we started looking around for anything that would work in the 900 MHz range as a high gain antenna. Then I happened to notice some old aluminum arrows that Daddy had in his "powder room" and it hit me like a ton of high explosives. We took a couple of the old arrows and cut them the right length to be a Yagi antenna like the kind you put on top of your house or RV to pick up local television channels. We drilled holes in a 1x2 board and slid the arrow antenna elements through the holes and then just glued them in place. The antenna worked well. It worked better when we spruced it up with a badminton racket handle and put a pistol-like grip on it. We had built, pretty much out of junk-nothing lying about the shop, a very nice and very directional 900 MHz communications antenna to receive video from our spy balloon.

The Russians have done stuff like this for decades. Their space program has been even more budget starved than ours but they manage to keep on going. After all, they *do* have the longest running track record with a heavy lift vehicle. And, we, brilliantly (that is sarcasm there if you didn't catch that) have outsourced our space program to them.

★ ★ ★

Americans aren't the only ones who have gone to space. Many other nations have their own space programs, plans, and astronauts. It is always a good idea to get good game tapes of the team you plan to play against in the upcoming game so that you can analyze them and make a good game plan for yourself. What follows are brief summaries of some of the other space programs around the world and how they might interact or impact or even supersede our present goals in space.

The Canadian Space Agency (CSA) was officially formed in 1989. Canada had performed some research into missiles and rockets ever since World War II, leading to the development of the Black Brant rocket (suborbital) in 1961 and the Alouette-1, their first satellite, in 1962 (launched by NASA).

The CSA has been internationally active from the outset, working with NASA and ESA in particular, and also with the Russians and

JAXA in the ISS age. They have their own cadre of astronauts, and built the robotic arm (Canadarm) for the Space Shuttle, and Canadarm 2 for the ISS, as well as the ISS Mobile Servicing System.

Their satellite activity continues apace, and in addition to continuing to participate on ISS, they plan a RADARSAT constellation of satellites (three) for Earth observation studies in 2014-15. The Polar Communication and Weather (PCW) satellites (two) are proposed to launch in 2016.

Unfortunately the agency is currently without guidance and has very little support in the Canadian government. The budget has flat-lined, and several promising programs, such as PCW, are without funding.

If they go to the Moon, it will be as an international partner with another space organization.

The Indian Space Research Organisation (ISRO) is the official space program of India.[8] With nineteen centers and contractors across the country, they are well positioned to take advantage of space.

India's space exploration began in the 1960s. Small sounding rockets carried experimental packages aloft to study the Earth's magnetic equator, which passes over India.

Since then they have expanded considerably. Kalpana Chawla was their first astronaut. She flew on STS-87 with the USMP-4 payload, as well as on the ill-fated final voyage of *Columbia*. They have placed several satellite constellations in orbit, including the Indian National Satellites (INSATs) for telecommunications, and the Indian Remote Sensing (IRS) satellites for observing and helping to maintain natural resources. Kalpana-1 is the first in a series of meterological satellites; geostationary and originally called METSAT when it was launched in 2002, it was renamed in honor of Ms. Chawla after the disaster that claimed her life.

The ISRO has two launch vehicles. The Polar Satellite Launch Vehicle (PSLV) places India's satellites into polar orbit, while the Geosynchronous Launch Vehicle (GSLV) carries satellites such as Kalpana-1 into a geostationary transfer orbit (GTO).

---

[8] http://www.isro.org/index.aspx

The PSLV is a four stage rocket, using both solid fueled and liquid fueled stages, alternating the fuel by stage, and can carry a 3,527-pound (1,600 kg) satellite into a 385-mile (620 km) sun-synchronous polar orbit, or a 2,315-pound (1,050 kg) satellite into a geosynchronous orbit. Its weight at launch is 650,000 pounds (295,000 kg) with a height of 144 ft (44 m). Its first launch in 1993 was unsuccessful, but since then it has launched nineteen consecutive times without serious problem.

The GSLV (Mk I & II) is a three-stage launch vehicle. The first stage is solid fueled with four liquid fueled strap-ons. The second stage is liquid fueled, and the third stage uses cryogenic liquid fuel. It will carry 4,400 to 5,500 pounds (2,000 to 2,500 kg) into GTO. At launch it weighs 913,000 pounds (414,000 kg) and stands 161 feet (49 m) tall, with a thrust of 1,478,000 pounds (6,573,000 N). Its first flight was in 2001, and it had five consecutive successful flights, but two flights in 2010 were unsuccessful. On one the GSLV-D3's cryogenic stage failed, and ISRO has not commented on the reason for the failure of GSLV-F06.

The GSLV Mark III is currently under development. It will be able to launch 9,920 to 11,000 lb (4,500 to 5,000 kg), 3,500 cubic feet (100 m³) payloads, enabling India to be not only self-sufficient in launch capability, but to compete in the global commercial launch vehicle market. A three stage vehicle, the first stage consists of two large solid rocket boosters strapped to the sides of the liquid fueled second stage, which is capable of restart. The third stage is the cryogenic LH2/LOX stage. Expected first launch is 2012.

Also in predevelopmental studies is a manned launch capability. The idea is desired to be able to place 2–3 crew members into a low Earth orbit and return them safely. A timetable for this has not been announced.

In addition, India has participated in many Earth, solar, and space science experiments, launched one unmanned lunar probe, and plans to send another in the not too distant future.

Unmanned lunar probes could be relatively easily followed by manned missions. The Indian space program is not one to be ignored. However, they are also cooperative, and most likely will work with other national programs on such grand goals.

★ ★ ★

There is much to be said, and insufficient room to say it, about the Japanese space program. It began in 1969 with the formation of NASDA, which became JAXA in 2003. NASDA successfully mounted the Spacelab-Japan mission on board a NASA shuttle, studying materials and life sciences in microgravity, and later Takao Doi became their first space-walking astronaut—for which he became a teen idol in Japan as famous as any rock star. Our American astronauts have for a long time lost the "rock star" status. We need to get that back.

The Japanese also built the Japanese Experiment Module (JEM) for the International Space Station. They have done considerable infrared and x-ray astronomy and very long baseline radio interferometry; studied the Sun and the Earth's magnetosphere and climate; and participated in developing communications satellite technologies.

Japan's first satellite was launched in 1970 with the L-4S rocket. Most of their early rockets were designs licensed from the United States. But in 1994, they introduced the H-II, the first all-Japanese rocket. The H-II has had mixed success, unfortunately, with numerous failures.

In 1985 they sent two probes to Comet Halley, Suisei and Sakigaki. In 1998 the Nozomi probe was sent to Mars, but failed to attain orbit around that planet. In 2003 they sent a lander, Hayabusa, to asteroid 25143 Itokawa, for a sample return. A successful landing was made, as well as return to Earth in 2010, but it is not known whether the sample collection and return worked. Hayabusa 2 is planned for launch in 2014 to a different asteroid. SELENE, their lunar orbiter probe, was launched in 2007 and its mission ended with a controlled lunar impact in 2009. In 2010 the Akatsuki Venusian probe was launched, but later that year failed to establish a Venusian orbit. BepiColombo is a joint venture with ESA to Mercury, planned to launch in 2014.

A total of five solar sails that use sunlight pressure for propulsion have been deployed by the Japanese space program—including the first, IKAROS in 2010—with mixed success. They hope to send a solar sail mission to Jupiter sometime in the 2010s.

The HOPE program was an attempt to create a Shuttle-type launch vehicle, begun in the 1980s, being budget-approved in 1988. The HOPE-X was to have been man-rated with a crew of four. In 1997 it was scaled back to a cargo craft to help augment shipments to the ISS, and is said to have looked quite a bit like the American X-20

Dyna-Soar concept. Unfortunately in 1998 it suffered a series of failures that resulted in a program review and substantial funding cutback, amounting to more than twice the amount that had currently been expended on the project from its inception. This slipped the first launch date back by over a year. In 1999, the standard H-2 lifter was canceled, and the H-2A initiated. This caused a further slide of a year in the launch schedule. Then NASDA was reorganized into JAXA, and the HOPE program was canceled entirely.

JAXA continues with spaceplane experimentation, but the designs are radically different from the HOPE-X, and there are no timetables for completion or launch.

They are a proactive partner in the ISS, and are likely to continue working with NASA and the US space program.

The European Space Agency (ESA) was formed in 1975, and is an organization formed from multiple governments dedicated to space exploration. Many European nations decided they could be more effective together in this endeavor than separately. The ESA is head-quartered in Paris, and has a budget of €3.99 billion ($US 5.65 billion as of 2011).

Its founding agencies, the European Launch Development Organisation (ELDO) and the European Space Research Organisation (ESRO), were born of the space race in the early 1960s when European countries realized two things: one, their scientists were slowly leaching out of Europe, mostly to America to work for NASA; and two, individually they did not have the power to compete with two superpowers, but working together, they might collectively accomplish similar achievements. ESRO managed to launch no fewer than seven research satellites, all in under a decade. But all of them were launched by the U.S., as they had no launch capability at that time. In 1975 the two organizations merged to become ESA. That same year, ESA launched its first major scientific mission, Cos-B, a gamma ray observatory developed under the auspices of ESRO. This occurred as the space race was winding down, and ESA emerged as a front-runner in unmanned space exploration.

The founding member states included Belgium, Denmark, France, Germany, Italy, the Netherlands, Spain, Sweden, Switzerland, and the United Kingdom.

By 1979 they had their own launch system. Ariane-1 was launched in that year, and Arianespace, the world's first commercial launch company, was founded in 1980 in France. Since then an entire family of Ariane rockets has been developed, including the heavy lift Ariane 5. The Ariane 6 is in development currently, and expected to enter service in the 2020s. Other launch vehicles used by ESA include Roscosmos' Soyuz-2 medium lift rocket and the largely Italian-developed light-payload Vega. Launch sites available to ESA include the Roscosmos sites as well as Arianespace's spaceport in French Guiana. The use of the Soyuz is by a joint venture agreement wherein ESA pays €340 million (US $450 million) to Roscosmos for the manufacture of Soyuz parts, which are then sent to the Arianespace launch site in French Guiana and assembled for launch.

ESA maintains its own unmanned space science missions, but works actively with the other major space programs: NASA, JAXA, Roscosmos, and ISRO, especially as regards the International Space Station, which is their principal manned space project. Whereas initially their main superpower partner was NASA and the U.S., changing circumstances in the 1990s, especially as regards the U.S. military, led to a realignment, and now ESA relies principally on Roscosmos and Russia for its manned access to space. More recently, several joint NASA/ESA projects were unilaterally canceled by the US and their budgets zeroed out.

Current member countries include:

- United Kingdom
- Ireland
- Spain
- Portugal
- France
- Belgium
- Italy
- Luxembourg
- Greece
- Germany
- Netherlands
- Switzerland
- Sweden
- Denmark

- ❀ Norway
- ❀ Austria
- ❀ Finland
- ❀ Czech Republic
- ❀ Romania

In addition, Canada is considered an associate member.

Together these member states contribute over 72 percent of ESA's budget. The European Union contributes an additional 21.6 percent. Cooperating states contribute most of the rest, and include:

- ❀ Hungary
- ❀ Poland
- ❀ Lithuania
- ❀ Estonia
- ❀ Ukraine
- ❀ Slovenia
- ❀ Latvia
- ❀ Slovakia
- ❀ Malta
- ❀ Cyprus
- ❀ Turkey
- ❀ Israel

Poland is, in fact, negotiating to become the twentieth formal member of ESA. The European Union's intent is to make ESA a formal agency of the EU; this has yet to occur, but is expected by 2014. Bulgaria, who had previously not been signatory to ESA but *is* a part of the EU, has agreed to join ESA.

In addition, the following nations have agreements with ESA, but are not members:

- ❀ Argentina
- ❀ Brazil
- ❀ China
- ❀ India
- ❀ Russia
- ❀ Japan
- ❀ USA

ESA has attempted semi-independent manned programs in the past, including the small shuttle Hermes, which would have carried three to five astronauts and 6,600-8,800 lb (3,000-4,000 kg).

Simultaneously they began developing plans for the Columbus Space Station. By the time planning was complete, the world political landscape had undergone a drastic change with the fall of the Soviet Union. ESA expected bigger, better things out of a cooperative deal with the Russians, and canceled Hermes and Columbus in 1995. So far, those bigger, better things have not come to pass, and the majority of ESA's manned space effort is involved in ISS.

However, programs since then have included Hopper, a German-ESA reusable LEO glider vehicle, launched aboard an expendable rocket. A prototype, the Phoenix, was built for a cost of €40 million (US $53 million). A remote guided drop test in Sweden in 2004 was successful. The project has apparently conducted no further tests.

Though not a manned mission, ESA has been a partner with the Chinese space agency (CNSA) in the Double Star mission, a two-satellite constellation designed to probe Earth's magnetosphere in detail. In addition, ESA, Roscosmos, and CNSA recently concluded the Mars500 simulated mission to the red planet, and are busily analyzing the resultant data.

An announcement was made that ESA would participate in Roscosmos' development of the Kliper, a glider design to replace the Soyuz capsule, able to carry up to six crew members, and to readily ferry crew and payload to and from ISS. After going through multiple designs, and offered as the complement to NASA's Orion Moon/Mars craft, a €50 million (US $66 million) participation study was not approved by ESA's member states and ESA dropped out of the program. Russia subsequently canceled the Kliper.

In 2011, ESA head Jean-Jacques Dordain was heard to remark at an airshow that Roscosmos and ESA would "carry out the first flight to Mars together."[9]

ESA is currently partnered with Roscosmos in a €21 million (US $28 million) study of the Advanced Crew Transportation System (ACTS), which would be a craft that would be capable of exceeding low Earth orbit. No stable design development appears forthcoming

---

[9] http://en.rian.ru/science/20110817/165853325.html,
   http://www.dailymail.co.uk/sciencetech/article-2027825/Russia-teams-European-Space-Agency-bid-launch-manned-mission-Mars.html

at this time, but matters may change in future. This agency is certainly in the realm of being able to attain exo-LEO manned flight, and has stated intentions to do so.

The history of the Russian space program is as long as that of NASA, and it has to be broken into two segments: the Soviet Union program and the post-Soviet program, as the funding differences between the two proved substantial and led the post-Soviet program in a rather different direction.

The Russian space program can be traced back to at least the early part of the twentieth century, when Konstantin Tsiolkovsky, a contemporary of America's Robert Goddard, proposed the notion that space flight was possible and began developing the modern science of rocketry. This was furthered by the Group for the Study of Reactive Motion, or "GIRD," as it worked out in Cyrillic. This group contained, among others, Sergei Korolev, who would eventually wind up becoming one of the Soviet's premier rocket scientists. In 1933 this Soviet team launched their first liquid fueled rocket, the GIRD-09, and a few months later, the hybrid fueled GIRD-X. It seemed Soviet rocketry was off to a great start.

It received a grave setback during the Stalin era, however. Stalin's Great Purge killed or imprisoned most rocket scientists of the day. It targeted the intelligentsia and professionals, among others, as presumably a threat to the Communist Party. Those who were not executed by firing squad were sent to the gulags. Many of the stories bear resemblances to atrocities being carried out at the same time but a little farther west. The only developments in the field of any significance occurred during World War II with the invention of the Katyusha multiple rocket launcher.

It was that "little farther west" that proved the reinvigoration of the Soviet space program, however. At the end of World War II, not all of Wernher von Braun's team ended up in the United States. He had purposely split his team in hopes that one or the other might make it to a safe haven with the Allies. The other group ended up with the Soviets, who looked over what von Braun's team had wrought at Peenemunde and were impressed.

Once in their new home in the Soviet Union, the German rocket scientists set to work helping the Soviets replicate the V-2. This

Russian version was dubbed the R-1. But it was not powerful enough to carry the large, heavy nuclear warheads of the day—which was the principal Soviet concern at the time. What the Russians wanted was a true intercontinental ballistic missile. That was always the whole point of their rocket program, as it turned out.

Nevertheless, the dreamers still got the dregs of the military's desire. Somehow having managed to survive internment in a gulag during the Purge, Korolev came out on top as one of the principal designers. But he had his enemies; the Soviet space program, unlike the American one, was not centralized, and there was much internal competition.

But in 1951, they succeeded in putting two dogs into a suborbital, non-exo-atmospheric rocket flight, retrieving them unharmed after ascending to 62.75 miles (101 km). When the United States announced, in 1955, their decision to put a satellite in orbit to commemorate the International Geophysical Year (1 July 1957–31 December 1958), Korolev was able to convince Khruschev, then in power, and his defense minister Dmitri Antoniou, to attempt to beat the Americans. And so in early October 1957, four months ahead of the Americans and coincident with the completion of the world's first intercontinental ballistic missile, the R-7 Semyorka—and not coincidentally launched by one—the Soviets placed Sputnik I in orbit around Earth.

One month later they launched Sputnik II. Sputnik II carried a living creature—a dog, Laika—into orbit. Laika did not survive her unique trip. Official reports claimed she was humanely euthanized at the end of the flight. Not until 2002 was it revealed that she was quite agitated and eventually died of overheating and stress only a few hours into the flight. The plan had truly been to euthanize her at the end of her voyage, but she didn't last long enough.

In short order, the USSR began making reasonably regular launches. Oddly enough, it has in recent years been determined that Khruschev had no particular interest in the space program. It was not a high political priority, he did not desire to compete with the Americans to the moon, and he saw its only benefit as propagandistic. Nevertheless the "propagandistic" program kept bringing home world firsts. In 1959, no less than three lunar probes were launched: Luna I, II, and III. These were, respectively, a flyby, an impactor, and a single slingshot orbit around the Moon that returned the first images of the lunar far side. And in 1960 they returned living dogs from Earth orbit.

But 1960 also saw possibly their greatest space disaster. It is commonly known as the Nedelin catastrophe after the manager who oversaw it. Some say he was cocky. Some say he ordered his subordinates into fatal position? It is likely that we will never know because there was only one survivor of this disaster.

In October 1960, a launch pad test for a prototype of the R-16 ICBM was in preparation. This test was overseen by Soviet Marshal Mitrofan Nedelin. The rocket was on the launch pad at the Baikonur Cosmodrome, and fully fueled with what they called "Devil's Venom," a particularly nasty hypergolic fuel system consisting of unsymmetrical dimethyhydrazine (UDMH) and nitric acid. The fuel mix is highly corrosive and toxic, but extremely powerful. However, even its exhaust is poisonous, and the nitric acid cannot be in the oxidizer tank for more than two days without eating through the tank.

It was the very nature of Devil's Venom, it seems, that led to the disaster. Late on the day before the test, the technicians preparing the rocket accidentally breached a line from the fuel tank, allowing a small amount of fuel into the combustion chamber. This in itself was not dangerous. So rather than go through weeks of untanking, repair, rebuilding the engine, and refitting, the decision was made to move up the launch to the next day. Meanwhile Nedelin notified many military and political dignitaries of the upcoming launch, in case they wanted to watch—and who doesn't want to watch a rocket launch?

Then on October 24th, engineers, scientists and technicians rushed to complete launch preparation, often performing tasks simultaneously rather than sequentially per checklist. The result was a delay. Nedelin, impatient and perhaps embarrassed, left the dignitaries at the observation stand and went to oversee the final preparation. Word has it he even brought a chair to sit down *beside the launch pad.*

For reasons unknown—or unreleased—the second stage engines fired prematurely. The flames acted as a blowtorch on the hypergolic tanks in the first stage, cutting through them. As soon as the UDMH and the nitric acid combined, as is the nature of hypergolic fuels, they self-ignited. The resulting explosion cremated everyone near the pad.

Those farther away were burned to death; those who had run away encountered the perimeter fence and were either burned or poisoned by the fumes. The entire horrific debacle was captured on film for posterity by automated cameras. Only one man survived—Mikhail Yangel, an associate (and competitor) of Korolev's, and that only because he left the viewing area and went into a bunker to smoke a cigarette. He had been the military's rocket engine designer and a proponent of hypergolic fuels. After this he was directed to concentrate strictly on ICBM design.

Meanwhile Korolyov was directed to focus on low Earth orbit with the Vostok and Voshkod spacecraft despite a desire to pursue the Soyuz design and lunar missions. But the Soviets suffered another setback in 1961. In a tragedy eerily similar to the Apollo 1 fire, but six years earlier, Valentin Bondarenko died in a high-oxygen (but low-pressure) environment during a training session. Yuri Gagarin kept vigil at his hospital bedside, where he died a few hours after extrication.

Shortly thereafter, aboard Vostok 1, that same Yuri Gagarin became the first human in space, completing a single full orbit of Earth. In the same year, the USSR launched Venera I to Venus, and put Gherman Titov into orbit for a full day aboard Vostok 2.

In 1962, the Soviets sent a probe to Mars, and launched Vostok 3 and 4, the first dual manned spaceflight. The next year, Valentina Tereshkova became the first woman in space. In 1964, Voshkod 1 carried the first three-man crew into orbit. In 1965, Voshkod 2 crewmen conducted the first extra-vehicular activity (EVA)— although not successfully. Their airlock was an inflatable, attached to the side of their craft, and it didn't work quite as well as envisioned. The Voshkod was a redesign of the Vostok, and not a particularly good one. It was cramped, it contained two crewmen instead of one without expanding the volume at all, and it had no provision for emergency escape. As if that weren't bad enough, after a little over twelve minutes of EVA, Alexei Leonov found that his spacesuit had ballooned out to a point where he had become inflexible. He simply did not have the strength to bend, even at the waist. When he attempted to reenter his vehicle, his suit became wedged and he couldn't reach anything to free himself!

In the end he had to vent atmosphere from his suit, risking the bends, in order to get it small enough to re-enter his spacecraft and rejoin his crewman Pavel Belyayev. *Then* the hatch wouldn't close properly. Once they got that fixed, the spacecraft was so cramped that, after orienting for deorbit burn, it took them an additional forty-six seconds to navigate their inflated spacesuits back into their seats. This threw off the center of gravity during the initial stages of reentry. The automatic landing system failed, and they had to resort to manual backup—*and* the orbital module didn't disconnect when it should have! They spun crazily until the module finally jettisoned at an altitude of only 62 miles (100 km).

The whole mess caused them to miss their designated landing area by a good 240 miles (386 km) in the middle of the Ural Mountains of Siberia. The location was cold, it was snow-covered, it was filled with bears and wolves—and it was the animals' mating season, the time when said bears and wolves were in their foulest moods. The Soviet control center had no idea where they were—and the hatch had been blown off by its own pyro bolts, leaving the capsule open to the elements. As was common in the early days of the space program for both Soviet and American, there was a pistol and ammunition aboard, and the men were trained for that terrain, but they had little in the way of shelter save the open Voskhod capsule. Aircraft located them, but it was too heavily forested for helicopters, so the men settled down for the night in their spacesuits—after stripping and wringing perspiration out of their soaked underwear. After a frigid -22°F (-30°C) night, a rescue party on skis arrived the next morning.

Not exactly a successful mission.

The Soviets' unmanned probes were having a little more success. From 1965-67, Luna 9 made the first soft landing on the Moon, Luna 10 was the first to achieve a lunar orbit, and Cosmos 186/188 managed a completely automated, unmanned rendezvous and docking.

Manned missions, not so good.

The N-1 rocket, the Soviet counterpart to the Saturn V, began development in 1965. Unfortunately its principal architect, Sergei Korolev, by this time known only by the enigmatic title "Chief Designer," as his very existence was a state secret, died abruptly in 1966 of medical reasons which are still debated. Cancer was certainly a factor, as was his known heart condition, but a botched operation, coupled with the inability to intubate him due to jaw damage from beatings dating from his days in the gulag, may well have contributed. This left the N-1 program leaderless. It floundered badly, and after four failed launch attempts, the program was suspended, then cancelled in 1976.

Soyuz 1 was the first flight of the Soyuz spacecraft. It was also the Soviets' first in-flight death. The craft was known to be faulty to begin with. The engineers reported over 203 design faults—not faulty equipment, not improperly installed, faulty *design work*, before the launch. Unfortunately, by this time Soviet leaders had caught Moon fever. They wanted to beat the Americans to a manned landing, and they wanted to take advantage of the delay caused by the Apollo 1 fire. Oh, and they wanted to celebrate Vladimir Lenin's birthday with some fireworks. Big fireworks.

Vladimir Komarov was the primary, and Yuri Gagarin was his backup. The situation was so bad that Gagarin tried to get Komarov bumped from the primary position, because he knew that he was considered a national hero and therefore not expendable. He hoped to get the mission delayed until the problems could be fixed. He failed.

Soyuz 1 was launched, Komarov aboard. Its mission was to rendezvous and EVA with Soyuz 2. As soon as it got on orbit, one of the solar panels failed to unfurl, so the spacecraft was running on low power from the get-go.

The Soyuz 2 crew prepped themselves for a repair mission. Thunderstorms overnight at Baikonur fried the Soyuz electrical systems, so Soyuz 1 was on its own.

Then the "orientation detectors" (I assume this means gyroscopes or star trackers or some such, or maybe not) decided to malfunction, rendering maneuvering difficult. Then the automatic maneuvering system died entirely, and the manual system went on the fritz.

Once the maneuvering system went down, the flight director decided to abort the mission. At this point, everything looked like a happy ending.

Except this was a new ship. With new details. Like a thicker heat shield, and a correspondingly larger parachute. Remember those design flaws? Guess what? Nobody bothered to make the chute receptacle any bigger. In their brilliance, technicians used wooden mallets to beat the parachute into place.

So the drogue chute came out, but the main parachute didn't. Simple enough: Komarov deployed the manual parachute. Which promptly tangled in the drogue chute.

He hit the ground at an estimated 89 mph (140 km/hr).

The ship exploded.

The Soviets didn't have too many manned firsts after that, and they never made it to the Moon with a crewed lander. The same year we landed on the Moon, they managed a docking and crew exchange of Soyuz 4 and 5. (The Soviets claimed that this was the world's first space station.) Unfortunately when it came time to come home, Soyuz 5's service module failed to separate, and the capsule with service module reentered nose first. The cosmonaut inside, Boris Volnyov, hung from his straps until the module's struts burned through and it broke away, enabling the capsule to right itself before the hatch also burned through—the gaskets were already burning and filling the cabin with noxious fumes. But then the parachute lines tangled, and the landing retros failed, and while Volnyov walked away from that landing, he broke his teeth. He landed in—yes, you guessed it—the Ural Mountains instead of Khazakhstan, and with the temperature outside at -36°F (-38°C), he was forced to walk several kilometers to the cabin of a local.

The next year saw an unmanned lunar soil sample return and the first unmanned lunar rover, as well as Venera 7's landing on Venus. The Venusian atmosphere is so corrosive and high-pressure that the lander didn't last long, though, but that was more or less anticipated. It is also believed the spacecraft may not have made a flat landing because of the terrain (which is invisible due to the opaque atmosphere until one is literally on top of it), and the antenna was therefore not optimally pointed for transmission to Earth.

In 1971 the Soviets put up the world's first space station, Salyut 1. Sort of like our Skylab, it was expendable and there was a whole series of these stations, military and nonmilitary. It was generally a successful program.

Except for the first flight to Salyut 1, Soyuz 11.

Soyuz 11 was the only manned mission to Salyut 1. All went nominally until it came time for reentry. At that time, the pyrotechnic bolts that were to release the service module from the reentry module fired simultaneously instead of sequentially. This in turn jolted open a breathing ventilation valve at an altitude of 104 miles (168 km) and bled the reentry vehicle's atmosphere off into space. As it was located underneath the seats, the cosmonauts couldn't locate and plug it fast enough to stop the loss of atmosphere. And due to the cramped conditions and the presence of three crew members, space suits were not worn for these early flights.

Flight recorder data later indicated the crew went into cardiac arrest within forty seconds. Within 212 seconds (less than four minutes) of the separation, the cabin pressure was zero. As a result, ground control lost communications with the crew long before the reentry comm blackout should have begun, realized that conditions were off-nominal, and began emergency preparations for the landing. The crew was found at the landing point, dead inside the cabin. Attempts were made to perform CPR by the service crew, but it was much too late.

In 1975 Soyuz 18a had the first ever manned launch abort. Its forward momentum carried it some thousands of miles downrange, nearly into China—which the Soviets were on particularly bad terms with at the time. It came down in the mountains again, sliding down the side of one, and nearly toppling off a cliff. This time, tangled parachutes saved the cosmonauts by snarling in the trees and preventing the sheer drop. The crew was pretty banged up.

In 1980 a Vostok rocket blew up on the launch pad. Forty-eight people died.

A satellite that had been launched in 1971 started orbital disintegration in 1981. There was a public outcry, fearing that there was a nuclear power pack aboard. At this, the Russians finally admitted that it was a prototype lunar cabin, indicating unequivocally that there had truly been a space race.

In 1986 the Soviets made another first when they put Mir, the world's first permanently manned space station, into orbit. With a lot of coaxing, it lasted long past its anticipated lifetime, and was finally abandoned shortly before it deorbited in 2001.

In 1988 they rolled out the Buran, their own version of a space shuttle, and amazingly like our own. There were rumors it was no coincidence, but as the Space Shuttle wasn't a classified program, there was no particular harm done. Like us, they built a fleet. But only one flew, unmanned, in 1988. After that, and with the USSR destabilizing, it was deemed too costly, and it was abandoned. The last anyone heard, the lone flight version was sitting near the end of a runway. It had been abandoned until, after the fall of the Soviet Union in 1991, some entrepreneur bought it, gutted it, and turned it into a space-themed restaurant.

This turned out to be a theme of the new, noncommunist Russian space program. Eternally plagued for money, Roscosmos, the new Russian space agency, turned to commercial endeavors, competing for commercial satellite launches, placing advertising on rockets, and inaugurating space tourism.

And they participated in construction of the International Space Station, of course, which is currently the longest-occupied "permanent" space station (if any station so far can be considered truly permanent). A Russian Soyuz took Expedition 1 to the ISS, and currently NASA purchases flights for our astronauts to and from ISS for a round-trip cost of $42 million per astronaut. For a time they also conducted space tourism to the ISS for a fee of $20-$30 million per person, but that has been curtailed due to the need for all ISS personnel to be functional crew.

Future Roscosmos plans include their own version of GPS, new science probes (Earth science, meterology, astronomy, returning probes to Mars and Venus), the development of a new launch vehicle, and, somewhere farther down the road when the ISS has exceeded its capabilities, a Mir-2. In addition, in partnership with ESA, they have stated the intent of developing the first manned Mars landing, a project that will almost certainly include Russians on the Moon.

China's space program as such began in the late 1950s, under the auspices of their Ministry of Aerospace Industry, and Chairman Mao Tzedong. At that time it consisted mostly of work on intercontinental ballistic missiles, as we were at the height of the Cold War, and they were responding to what they considered potential threats from both the U.S. and Russia. They seemed to have no particular interest in manned space flight for several more decades.

Upon Mao's death in 1976, Deng Xiaoping emerged as China's leader, and canceled many missile programs and anti-missile defense programs considered important at the time. However, long range ICBM development did continue, as well as the Long March series of launch vehicles, enabling them to compete in the commercial launch industry. When the Cold War ended, Deng stepped up his commercialization of China, and moved away from the blatant use of communist revolution rhetoric in the naming of vehicles, and toward ancient Chinese religious and mystical names. This included, for example, renaming the Long March rockets "Divine Arrow."

He split the Ministry into two parts in 1993: the China National Space Administration (CNSA), responsible for space policy and planning, and the China Aerospace Corporation (CASC), responsible for execution of the program.

Shortly thereafter, China had its first public space program disaster.

In February of 1996, the launch of the first Long March 3B heavy launch vehicle went drastically wrong. Carrying Intelsat 708, a commercial telecommunications satellite, the rocket failed almost immediately on liftoff as a result of an engineering defect, deviating drastically from its launch trajectory at the Xichang Satellite Launch Center. It crashed twenty-two seconds later and slightly more than one mile (slightly under two kilometers) from the launch facility— directly on top of a village. Xinhua, the official Chinese news agency, under government control, reported six killed and fifty-seven injured, with eighty houses destroyed. Unofficial reports, however, place the death toll at well over 500 people.

Three years after this disaster, Shenzhou 1 was successfully launched—unmanned—on the anniversary of the founding of the People's Rebublic of China in 1999.

Only ten years after the Ministry was split, and perhaps pointing to something positive about the decision to do so, Shenzhou 5 carried the first taikonaut into space on October 15, 2003.

Since then the Chinese have been very busy planning and executing an ambitious program of space development. They already have six "space cities" and four launch sites, and are building more. Their first successful spacewalk occurred in 2008; in October 2011 they launched their first laboratory module, Tiangong-1, and in November completed a successful docking and return of Shenzhou-8 with it. This is a huge step toward a permanent orbital space station.

The Long March family of launch vehicles is having three new members added to its list. These will be heavy lift vehicles using a new, "non-toxic" fuel, although that fuel is not specified. Long March 5 will carry 55,000 pounds (25,000 kg) to low Earth orbit, and 30,800 lb (14,000 kg) to geosynchronous orbit. Long March 6 will carry 2,200 pounds (1,000 kg) into a sun-synchronous orbit, and the Long March 7 will carry 29,700 pounds (13,500 kg) into low Earth orbit or 12,100 pounds (5,500 kg) into a Sun synchronous orbit.

The Chinese keep the budget of their space program a closely guarded state secret, partly because it is mostly military in nature. They do promote it as a diplomatic and economic boon. This is mainly propagandistic, as much of the early Soviet space program releases were. Currently, their budget is claimed to be something over US $2 billion annually, which is a fraction of the NASA budget, even as small as *that* is. It is important to realize two things, however: 1) much of their near-term spacefaring involves unmanned probes and the equivalent of our Projects Mercury and Gemini. 2) This may be only the "civilian" part of the budget. 3) While it sounds very heartless and maybe cynical, it is a fact that, due to cultural imperatives, the Chinese have a decided surplus of single, unmarriageable men and a distinct lack of marriageable women. Given the official attitude toward the Long March disaster, it may be that they feel that they can afford to be less scrupulous regarding safety measures, which in itself is a cost-saving measure.

Short-term, they are endeavoring to develop and place an Earth-observing system, a telecommunications network exclusive to China, competitive commercial launch services, orbital studies/satellites/telescopes (for astronomical, physical, life, materials, and microgravity sciences), their own version of the Global Positioning System, and unmanned lunar probe exploration plans.

In the longer term, things get more complicated. This seems to be, effectively, their very late entry into the Space Race, but it is one they intend to win. It's a goal which may be possible, as we have essentially ceded the field. Their future plans include:

- ❀ A manned space station. This would be a modular concept, like the International Space Station, but built and coordinated and overseen by only one nation, as Space Station Freedom would have been. They expect it to be complete before 2020 and to weigh 60 tons.
- ❀ Their own manned lunar landing series. They are already

intensely working on intermediate to heavy launch vehicles capable of placing 50 and 500 tons, respectively, into a trans-lunar injection trajectory. In 2008, Chinese television "let slip" images of the Chinese lunar rover, which may be already developed. They expect their first manned landing (following on from the unmanned lunar probes) by 2025.

❀ They intend to construct a lunar "observatory" and base, which would be a permanent fixture with a continuous manned presence. One of their space leaders has indicated this was a vital step in their plans to push further outward.

❀ To that end, they will set up a space weather forecasting system at the Sun/Earth L1 point, to determine when solar flares and coronal mass ejections may pose a problem for taikonauts outside the Van Allen belts.

❀ They will begin unmanned Mars probes in 2014 and continue until 2033.

❀ They will initiate the first manned Mars mission in 2040 and continue until at least 2060.

China is a member of the UN Committee on the Peaceful Uses of Outer Space. However, its space program, despite the "corporate" designation of half of it, is entirely military-run, and in 2007 it shot down one of its own dead satellites.

With a high technology space plan and national mandate like the Chinese plan it is clear that, unless challenged, China WILL be the high technology super power in less than two decades from now. Americans have to realize that the Chinese political system will allow for long-term planning and goals and in fact such methods of planning have been in place for centuries or longer there. Our election-to-election thinking and lack of planning is why we have lost our foothold in space. We should look at this Chinese plan as a slap in the face and use it as a kick in the pants to wake us up and move us back onto the path of rebuilding our own technological superiority. Americans should not give space away to the rest of the world. We should at least be racing with them every step of the way. I would prefer us to be leading the way.

My good friend Stephanie Osborn was a payload flight controller for both Shuttle and Station, and she's had some experience with a number of different national space agencies besides NASA. I talked to her recently about her direct experience with them, and this is what she had to say.

"My work as a payload flight controller at the Marshall Space Flight Center has enabled me to work with almost every space program in the world—ESA, NASDA, India, and Russia/Ukraine. (China excepted; they don't seem to want to play with others.) It's been fascinating to meet so many people from around the world, and it's been equally fascinating (and sometimes scary) to see how the other programs operate.

"My earliest introduction to the Russian program was when they had a launch during one of our Shuttle missions. (I couldn't tell you anymore if it was manned or unmanned, to be honest; it's been awhile. I *think* it was unmanned. I'd like to think a manned craft would have a little more . . . complex . . . launch system.) The Payload Control Center piped in the video on one of the closed circuit TV loops, and I watched the last moments of what we would consider the countdown.

"The first thing I saw was a big old hangar, looked like it was around World War II vintage, surrounded by tall grass, probably hip high. Poking up out of this tall grass was the occasional piece of old equipment—you could tell it was old because the only color any of the pieces had was solid rust. And they were scattered around in such a way that it looked like somebody had ordered, "Get rid of it," and the person in charge of getting rid of it just heaved it out the door of the hangar and there it sat wherever it landed.

"A train track led out of the hangar doors, and shortly a diesel engine chugged out. Behind it, on big ol' flatbeds, the rocket lay horizontally. It was escorted by great big military officers in decorated heavy coats and the big fur hats they wear in the cold. These escorts were on foot. Walking. Almost but not quite marching. That's how fast—or slow, rather—the train carrying the rocket was moving. I guess it was comparable to our big crawler transport and crawlerway that runs from the Vehicle Assembly Building to the pad. It doesn't go very fast either. But I don't think we have military guys walking alongside it to provide a formal escort.

"So the train trundles out a ways from the hangar and then stops in the middle of a field. Some sort of special crane was used to slowly raise the rocket into a vertical position; it looked sort of like a portable version of the gantries we use. The train chugged its way out of view of the camera, and final checks were performed.

"NASA astronauts who have flown with the Russians tell me that, at this point, the bus containing the flight crew arrives and they all get out and urinate on the bus tire. It's a specific tire, too, but I'm not sure which; one of the front ones, I think. This is supposed to be good luck, and all the crew members have to whizz on the tire. I don't know how they work that out for the ladies. Then again I'm not sure any female cosmonauts or astronauts have flown on a Soyuz in recent years. But after that, they're put inside the capsule, pretty much the way we do it over here.

"So anyway, back to the live video feed. Once everything had been rendered flight-ready and the ground crew had safely cleared the area, the rocket was launched. Right there, in the field. No special pad, no fire suppression system, nothing like that. Once it was apparent that the rocket was well on its way and wasn't likely to fall back on top of the launch site, a swarm of fire engines came in and put out the grass fire in the field. And that was the launch.

"Some astronauts that I knew had been on Mir, and I heard it was kind of strange. If one of the cosmonauts came across something in stowage he couldn't immediately identify, he called the ground. If an answer was not soon forthcoming, said unidentified something got chucked out the airlock. Literally. Supposedly the rationale was, if it can't be identified instantly, it must not be important.

"And I heard about the astronaut who was on Mir during the fire, from another astronaut, a mutual friend. The fire was caused by a malfunctioning oxygen generator. The Russians downplayed the incident, but it seems it was more serious than they let on, because I had it to understand that the astronaut literally couldn't see his hand in front of his face for the dense smoke, and the crew was very close to abandoning ship when they finally got matters under control.

"One Shuttle mission I worked on was going to have a Ukranian astronaut, and NASA Headquarters was looking at manifesting a Ukranian welding experiment in order to check the system out for possible upgrade and use on ISS construction. At that time there was a neutral buoyancy tank at MSFC and I was involved in the training and general checking-over of this experiment for flight. The intent was to make sure it was compatible, and then waive the safety cert by listing it as a reflight, since it had already been up in the Russian program at least once.

"I was intrigued. I didn't know much about welding at the time, so I went out and got some books and read through them when I was in my office. And the more I learned, the more worried I got.

"You see, this was basically a form of arc welder, and it produced an electron beam from a gun grip, which was then directed at the object(s) to be welded. Different guns produced different welding effects that might be desired—cutting, melding, etc. It was going to be carried in the payload bay of the shuttle, necessitating an EVA to operate.

"For eye protection, the sun shield visor of the space helmet would be sufficient. But the guns had no hand protection in the form of a guard; there was no adjustment for the grip if the EVA crewman (or woman) did not have a hand that fit the grip. There was also no height adjustment—if you couldn't see what you were doing from the foot restraints provided, tough luck. This alone limited the usability of the experiment to only certain crewmen.

"There was no particular protection from molten metal spatter that I could note, save for the console of the welding apparatus itself. Worst of all, there was no safety mechanism to cut off the device's trigger if an accident occurred, or the astronaut/cosmonaut unthinkingly moved the beam gun into a dangerous angle while trying to adjust something else. (I mean, c'mon, who hasn't spilled their coffee trying to hold it and look at something else simultaneously?) That way led to suit breaches, slashed gauntlets, and the like.

"Yet it had been flown by the Russians for the Ukranians. This made about as much safety sense to me as the human bucket-of-cement brigade that had been used at Chernobyl. In the end, it was not manifested, much to many NASA folks' relief after seeing the thing.

"There were two Ukranian candidates to fly that mission. Both were Ukranian air force fighter pilots, one a colonel, the other a major, if memory serves. Both were considered the cream of the crop—the best candidates the Ukraine had to offer.

"But before they could pass their flight physicals, both would have to have extensive dental work. Note I did not say as an adjunct to their physicals—they could not *pass* their physicals until this was done. Why?

"Whether or not the welding experiment flew, there were going to be EVAs—space walks—on this mission. Space suits have fairly low pressure, despite the difficulty in moving in them. So the shuttle cabin would experience a preparation period of gradual depressurization.

"Unfortunately, cavities in teeth are porous. They *will* hold atmosphere, but they are not sufficiently porous to release it very quickly. Cabin depressurization would cause severe pain, and potentially even shatter the tooth involved. Therefore these men, the very best their country had to offer, military men whose service physicians should have been taking care of such things, were looking at extensive dental work before they could even think of flying the mission—extensive as in, I heard the words "crown" and "bridge" tossed around. I felt bad for them.

"Rumor in the space community also had it that the Russian heat shielding, at least early on (and I do hope they've become a bit more sophisticated about it now) consisted of a large block of wood carved into an aerobrake shape and attached to the bottom of the re-entry module. It then ablated (burned away) in the friction of reentry and protected the crew within. This is, by the way, also rumored to be the route that the Chinese are currently taking, using specially treated cork."

Now, I don't know about you, but it sounds to me like we've got a better thing going.

I wanted to add Stephanie's input here to give a flavor of how some of the rest of the world does business in space. Although Steph and NASA didn't like the Ukrainian welding experiment because it was a "safety nightmare," I do. I think that is part of what has been lost with the American space program is the "derring-do". All we needed was to jump in and roll up our sleeves and figure out how to make the welding experiment a little safer and more useable and then do it. We need that capability in space along with a million other construction capabilities. I will say that I understand and agree with Steph and NASA's need for safety and requirement that every flight and experiment is as safe as possible. But there also must be a way to push the daring edge a little more while not being too risky. Perhaps the experiment should have required at least a couple of astronauts working it, and additional modification of equipment. The problem with those "modifications" is that it takes it out of the realm of a reflight and you have to pass safety inspections all over again, which is what they wanted to avoid, to streamline the process. When we weld things on the show there is a requirement of a minimum of two men. Somebody welds and somebody is on safety and fire watch, always. I'm sure there were procedures being planned along these lines.

In addition, it may be important to note that, although there is an international Agreement Governing the Activities of States on the Moon and Other Celestial Bodies, otherwise known as the Moon Treaty, it is considered a failed treaty, as no spacefaring nation— including China—has ratified it. The only signatory nations are those who have no space programs of any sort.

This raises concerns that a territorial dispute could result should another nation construct a lunar (or Martian) base and claim ownership by right of occupation.

CHAPTER 6:
IF THE SATURN V IS GONE, HOW DO WE GET THERE?

I recall designing and building my first rocket like it was yesterday. I was six years old and it was summer vacation after the first grade. That puts it in late summer after July and in 1974. I had been watching rockets, *Star Trek*, and all sorts of other science fiction shows on television. My brother and I could hear and see the rocket motor tests across the river on the base and I wanted to fly one myself.

I gathered up what parts I could find that might work as rocket parts. I specifically remember an empty paper towel roll is what triggered my enthusiasm. Then I found some cardboard from a macaroni-and-cheese box and I glued fins on the thing. I cut up a garbage bag and used some kite string to make a parachute. But building a nose cone was quite an issue for a six-year-old! I finally convinced my mom to give me the plastic egg-shaped container from her Leggs pantyhose. It was a little big but it worked.

When my dad saw me reading in our Science Encyclopedias how to build a rocket motor he got a bit concerned. Daddy has always been a nervous Nelly. So, he bought me an Estes rocket motor instead. I remember it was a C engine. It was too small in diameter to just shove up in the paper towel roll. I learned my first lesson about the "thrust structure." I wrapped paper tape around the engine until it would barely fit up in the tube. I shoved a pencil through the tube just ahead of the engine, broke off the excess, and then wrapped some tape around that.

My big brother and I went out in the yard and set the rocket off. I learned then that cardboard, flimsy mac-and-cheese-box cardboard, was useless as aerodynamic structures. The fins flapped like birdwings and the rocket did loop-d-loops and chased us across the yard. Learned a bit about Safety that day.

I was enamored with rockets, fireworks, and things that went boom ever since that day, however. When we filmed the 4th of July episode in the first season of the show I was like a kid in a candy store with a pocket full of money. Well, more like, I was a redneck kid in a fireworks store with a pocket full of money!

The competition the guys and I had to see who could make a rocket out of junk in thirty minutes was a lot of fun. Rog knew to get on my team because he had the benefit of seeing me fly rockets in school. Michael, my nephew, should know how to build a rocket. I showed him how to build them when he was six and took him out to launch them a few different times.

But the competition was tilted in my favor. After all, I did write a textbook on how to build rockets. The key to our rocket, to any rocket for that matter, is to follow the design rules. Rockets have certain parts: a nose cone, a body tube, a thrust structure, fins, and sometimes a recovery system (chute). The key rule is that the center of gravity must be higher up the rocket than the center of pressure for stable flight. This means that the rocket needs to be top heavy so it is always wanting to fall over. The pressure from the drag on the fins pushes the bottom back down and the two opposing forces keep the rocket stable. If the center of gravity is too low the rocket will tumble.

Pete and Michael built their rocket out of crutches with the big heavy parts on bottom. And, they didn't put a thrust structure on it. So when they lit their engine the rocket fell over and the engine broke free and flew across the grass starting a fire. Spectacular, but it didn't fly.

Mine and Rog's rocket used a toilet bowl seat for fins, a minikeg for a nose cone, and we ran bolts through the body to hold the engine in. Our rocket flew well until one of the fins pulled free and flopped about just like my first cardboard rocket when I was a kid. It barely missed a parked car out the field where we had warned people not to park.

★ ★ ★

Watching the fireworks episode or the moonshine rocket, or any other episodes, you might notice that accidents *do* happen but we have yet to get more than scratched or lightly singed or just shaken up a bit and no first aid really needed more than just requiring us to change underwear. This is because safety is always on our minds. Daddy has beat safety into all of us since we were kids. And he still does it every weekend. This is why I push the guys to recall Safety 3rd!

Safety 3rd came about while we were working on the Moon buggy. Rog was cutting some steel and Daddy told him a certain way to use one of the saws and told him he'd end up like a three-fingered shop teacher if he didn't use the saw right. Rog held up three fingers and cracked the joke, "Safety third."

I then realized that Safety 3rd was a very fun way to keep safety on the minds of the guys. It was funny and it gets across the warning that if you let safety be third then you'll end up maimed or worse. So, we put together some rules that we go by. Actually, one of our production assistants, "Sloth" as we call him, was first to say there should be three rules. So we developed the following three rules:

Rule #1: Always follow Rule #3
Rule #2: When in doubt refer to Rule #1
Rule #3: Safety!

When you know what you are doing, such as Daddy, who is a master toolmaker and OSHA-trained safety man, does when he is building a tool or working with his drill press, there is almost no apprehension, and mistakes become part of the expected process and not something out of the blue. He plans for the mistake so that if/when it does happen there is no mishap, only a bit of frustration and lost time. My fellow Rocket City Rednecks are competent in what they do, even if sometimes we poke fun at it and their very good skillsets aren't obvious from what gets on television. Competence, such as Daddy's and Rog's basic mechanical technical building skills, Michael's welding and electrical circuitry knowledge, and Dr. Pete's general knowledge of all things physics provide a good lesson for safety in our redneck science and space exploration. How so?

Because we have practiced our particular skill set every day of our lives until we are really good at them. It is why astronauts drill over and over checklists and mission scenarios so that they aren't startled by any mistakes or malfunctions. Instead they are well trained and expect such to happen and know what to do and how to handle it.

This reminds of a day when one of Michael's buddies stopped by Daddy's garage while we were filming. He had stopped by on an old tractor that he had repaired and wanted to show us. He left it running sitting there in the driveway and went in to talk to the guys. I had been in the restroom and was walking out when I heard people screaming that the tractor was loose. Cameramen were running around getting out of the way and people were pointing as the tractor headed for the garage at a pretty good clip of a few miles per hour. Rather than screaming and pointing I simply walked over, climbed up on the tractor, and stepped on the brake. Once Michael's buddy came back out and turned off the tractor I did low-rate the heck out of him for not setting the emergency brake. Checklists! It is all about checklists!

We see all the time that "Whoa! We need to add a checklist before we do this again."

Just like any good space mission, there should always be a checklist. As we develop new rocket designs. Checklist. Once we build nuclear rockets. Checklists. Solar Sails might fly. Checklist first. Even if we build a warp drive (I should say *WHEN* we build a warp drive) we will go through the Checklist before we do it. Safety third!

★ ★ ★

We've already talked about the Saturn V and how it worked, how awesome it was, and why we can no longer build them. We could go get all the information we have on them, find the pieces of them that still exist, and do a massive reverse engineering effort. But why would we want to reverse engineer sixty-year-old technology? Remember that all of the Apollo era stuff was built without modern computers and construction methods. We should learn from Apollo, but not redo it. We should build a new generation of launch capabilities and space vehicles with modern and next generation technologies to push us forward rather than take us back to our glory days. Our glory days should still be ahead of us.

We recently retired our other main historic heavy lift system, the

Space Transportation System, commonly known as the Space Shuttle. The origins of the Space Shuttle actually date back to the late '40s and '50s of the previous century, in the X program of hypersonic rocket-powered aircraft, jointly sponsored by NACA, the USAF, and the U.S. Army. The very first X craft, the Bell X-1, dropped from the bomb bay of a B-29 and piloted by Captain Charles "Chuck" Yeager, broke Mach for the first time, achieving a peak speed of Mach 1.06, or 807.2 mph (1,299 km/h).

The X-15 (late '50s through '60s) in particular set many speed records (highest speed Mach 6.7 in 1967), as well as an altitude record of sixty-seven miles (108 km) in 1963. There were plans for an X-20, aka Dyna-Soar, which would have made an exoatmospheric (space) flight. The test pilot selected for that flight was none other than one Neil Armstrong. But the program was canceled before the X-20 was built. A later version, called the HL-10, was delivered to NASA in 1966.

The X-15 was, typically, slung under the wing of a B-52 bomber and carried to an altitude of about 45,000 feet (13,700 m), and while moving with the B-52 at speeds around 500 mph (805 km/h), was dropped from the bomber. Fractions of a second later, the rocket ignited, and propelled the X craft at high speed.

Some considered the pilots of these vehicles America's first astronauts, and they were, in fact, awarded special wings, per some reports. There is some debate about whether or not they actually did go exoatmospheric. The test pilots reported atmospheric interface effects as they achieved maximum altitude and started back down. This, they found, could be tricky, as, if not done at exactly the right angle and speed, there was a tendency for the effects to place the craft into a flat spin, which is next to impossible to recover from.

Certainly all of the information learned in these flights aided NASA in projects Mercury, Gemini, Apollo . . . and the Space Shuttle.

First, let's talk about the actual shuttles that flew, the ones that were just retired.

The Space Shuttles, as flown, were stacked up on the launch pad along with one very large External Tank (ET) housing cryogenic liquid hydrogen and liquid oxygen for fuel and oxidizer respectively, two large Solid Rocket Boosters (SRBs), and one orbiter, aka space shuttle, as well as the various and sundry structures connecting them. Overall height of the stack was 184.2 feet (56.1 m). It had a mass of 4.47 million pounds (2.03 million kg), and its diameter . . . well, that's a bit difficult to determine for such an asymmetric stack, but it was about 28.5 feet (8.7 m) across. It had a payload capacity to low Earth orbit of 53,600 pounds (24,400 kg); to geosynchronous transfer orbit of 8,390 pounds (3,810 kg); to polar orbit of 28,000 pounds (12,700 kg); and a return-to-Earth payload capacity of 32,000 pounds (14,400 kg). It had a total of 135 launches, 134 successful launches, and 133 successful re-entries. It had a total combined thrust (both SRBs, all three SSMEs) of over 6.8 million lb (30.5 million newtons).

In addition, it had the Orbital Maneuvering System (OMS), consisting of two pods on either side of the vertical stabilizer, containing one AJ10-190 hypergolic engine each. (Hypergolic fuel is typically a hydrazine, in this case monomethylhydrazine or MMH, and when combined with the appropriate oxidizer—usually dinitrogen tetroxide—spontaneously ignites, requiring no mechanical igniter. It is toxic, but produces substantial thrust.) These were used to provide supplementary thrust on launch, large maneuvers on orbit, and the deorbit burn. They had individual thrusts of 6,000 pounds (26,700 N). Smaller hypergolic fuel jets called the Reaction Control System (RCS) provided more delicate manuevering on orbit. When you would hear the astronauts say, "Preparing for OMS burn" they weren't doing yoga exercises, they were firing the Orbital Maneuvering System.

Initial shuttle flights had the nomenclature of SSP-#, for Space Shuttle Program Number whatever, but quickly the designation was changed to STS, for Space Transportation System, and this is what most of us are familiar with.

The original plans (and there were many different concepts looked at before anything was settled) called for a craft much more like the X series, with both liquid rockets and cryogenic fuel tanks located within one or two fuselages, and behind the crew compartment. It was planned to launch in a vertical position, unlike the X planes, and glide back to a runway landing just as the X planes did, and as the final design ended up doing.

But at the time this design process was occuring, another process was occuring on Capitol Hill. That process involved determining what the post-Apollo space program would do. Options were:

- A manned mission to Mars program
- A return to the Moon for follow-up studies (maybe a base?)
- A low Earth orbit program
- Cancellation.

President Nixon and his advisors decided not to cancel a successful program at its pinnacle, but at the same time did not have the . . . confidence, forward vision, courage—call it what you like . . . to choose one of the more advanced ideas. It is more likely that Nixon's administration simply did not want whatever Nixon planned to look like the Kennedy plan. So, they made the call for a low Earth orbit program, and there we've been, ever since. No administration since has had the gumption to push Americans back out into deeper space to boldly go and do daring stuff.

Discovering there wasn't enough in the budget for a craft like they originally wanted, and that they could increase the size of the payload bay with a redesign, the designers compromised and created an external fuel tank for the shuttle. This ET is really an amazing design in itself, as it uses the pressure of its contents as part of its structural integrity. This maintains a light weight, but provides the strength to endure the stresses of launch. This type of construction is called *monocoque* construction and there have been several rockets and missiles throughout history like this.

Then they found there still wasn't enough money to build the kind of powerful engines they wanted. So the designers compromised again, scaled back the size (hence thrust) of the Space Shuttle Main Engines (SSMEs), and designed a liquid fueled external booster.

Once again budgetary constraints stepped in, so they ditched the liquid external booster and strapped on two large solid fueled rockets. This became the final configuration, despite the fact that engineers here at Huntsville were extremely worried about using solid rockets on manned missions. Unfortunately, as we have since found out to our sorrow, the Marshall engineers were mostly right, and it is far from the best configuration. The problem isn't necessarily the SRBs themselves but the combination of placing the SRBs, the large ET full of fuel and oxidizer, and the Orbiter full of astronauts all right on top of each other around the fire. Thrust is the problem and was eventually what became an issue in both the *Challenger* and *Columbia* disasters. The astronauts need to be above the fire and volatiles so their crew cabin is free of explosions and flying debris. With all of the components together on the same stage level like the Shuttle is known as *parallel staging*. Staging like the Saturn V is called *serial staging*. Serial staging is inherently safer because all the stuff that can go boom or shake free and hit anything is below the crew cabin.

The shuttle's solid rockets have two main difficulties when being used for a manned launch. The first is that they are basically sophisticated bottle rockets—once ignited, they cannot be turned off. The second is that the solid fuel is cast, like concrete or really more like rubber, inside the skin of the booster (the casing), with a central hollow core in the shape of a multipointed star pattern called the *perforation*. It is along the perforation where the fuel is ignited (it burns from the inside out, not from the bottom up like a hobby rocket motor). And, like concrete, there is only so big a size that cast can be before it will crack during curing. Now, cracks in concrete may not be so bad, depending on what the concrete's used for—but cracks in solid fuels are disastrous, because the crack provides a pathway for the combustion to follow, resulting in uneven thrust and in fact burnthrough of the casing, if the combustion reaches that far. This means that something as large as an SRB has to be cast in segments, then assembled into one large booster. O-rings (yes, the infamous O-rings) are used between the segments to help provide seals between them. The O-rings really aren't bad guys here. In fact, if the rings burn through they typically weld the casing back shut in seconds. The problem is putting a fuel tank right beside them or a mooring point to the rest of the rocket right by one of these segment points. It is believed that during *Challenger* that the O-ring burned through at the bottom, which allowed a flash of rocket exhaust (think blow torch cutter) to hit right on top of the mooring structure holding the bottom of the booster to the rest of the Shuttle. Video shows the O-rings sealing back off and the flight going fine until about sixty seconds later when the weakened mooring structure finally broke free. This was a slight design flaw with hindsight and had we not ever operated the Shuttle in freezing conditions outside of the appropriate operating temperature of the O-rings we might never have discovered it. I hope we learned from *Challenger*.

There are two major issues which mark the Shuttle program as different from all previous NASA manned programs in safety. These differences are due to politics and to money. Politics cut the money to the point that these design choices had to be made.

- ❧ The Shuttle configuration is the first major lifting vehicle that does not have the capacity to achieve a stable orbit with the loss of one engine.
- ❧ The Shuttle configuration is the first manned spacecraft that does not have a means of detaching the crew compartment in the event of a catastrophic malfunction.

Point 1) above is why there are staged checklists between launch and orbit, and why multiple sites must be "go" before a Shuttle launch. These multiple sites are designated mission abort landing sites in the event of engine failure. Abort modes include:

Return To Launch Site (RTLS)—an RTLS abort would occur shortly after launch. It requires the jettisoning of both SRBs and the ET, and the purging of fuel and oxidizer from the SSMEs. The orbiter must then right itself (it goes into orbit in an upside-down attitude) and bank around 180°, then line up with the runway at the Cape and land—all while in glide mode. It is completely unknown whether this abort mode is even possible, but it is not considered very likely. More probable is the jettison of the SRBs and ET, righting the orbiter, and executing an emergency bailout, similar to what was seen in the movie *Space Cowboys*. This abort option did not even exist prior to the *Challenger* disaster. And it would prove to be difficult to accomplish in a short amount of time.

East Coast Abort Landings (ECAL)—these were added after *Challenger* as well, and after regular high-inclination flights to ISS were instituted. These are similar to an RTLS abort, but the emergency landing sites are along the East Coast. They are not much more favored than an RTLS, and for the same reasons. In addition, many of the emergency sites are in heavily populated regions (such as Atlantic City and the JFK airports) and would have little to no time to prepare for the landing. With the overcrowded and overused status of our nation's current air travel network these abort sites are less than optimum.

Trans-Atlantic Abort (TAL)—this would occur with an engine failure a bit farther along, while the Shuttle was still over the western

Atlantic. Again, the ET and SRBs would be jettisoned, the orbiter would be righted, and it would make for a landing site in either Spain or North Africa.

Abort Once Around (AOA)—occurs when the Shuttle has too much momentum to make the TAL sites. The probability here is that the SRBs have already spent and dropped away, and the failure has occurred in a SSME. There are multiple sites around the world prepped for this possibility. The runway at Fort Campbell Army Base, Kentucky, was at one time an example, as it must be an extremely long runway to account for the Shuttle's high initial speed.

Abort To Orbit (ATO)—this is the one type of abort that has actually occurred. This involves engine failure at an extremely late point in the orbital insertion, where achievement of the expected orbit is not possible, but a lower orbit is attainable. It cannot be maintained long, due to atmospheric drag, however, and so essentially all that can be done is to deploy the minimum equipment needed to function and stay in communication with the ground, and prepare for a deorbit within a couple of days. This occurred on STS-51F due to a SSME failure, and very nearly on STS-93 due to a fuel leak in one of the SSMEs.

There are approximately seventy-five emergency landing sites for the shuttle, scattered around the world, on every continent except Antarctica. An abort there might scare some penguins out of their little minds anyway. And no telling how long it would take to get to the astronauts after the landing.

The Space Shuttle does have one distinct advantage over other manned vehicles: crew complement. The typical crew consists of seven astronauts, although eight have been flown on some missions. The orbiter can hold up to ten or more astronauts, however, more than any other vehicle flown or currently envisioned.

One of the reasons why the Shuttle never achieved its hoped-for rate of launch was directly due to the final design configuration, which was in turn determined by budget (a.k.a. politics): NASA's Michoud Assembly Facility could only build twenty-four external tanks per year. Another reason was that Station and USAF missions were insufficient to keep it as busy as intended. As a result, other boosters were phased out, and the Shuttle became the only vehicle to place most American payloads into orbit. This kept it busier, but still not busy enough, and the cost per pound to orbit proved prohibitive for many experimental programs' budgets.

The Space Shuttle fleet, now retired, consisted of a maximum of four flight-worthy vehicles at any one given time. *Enterprise* was the test glider. The initial flight-worthy fleet consisted of *Columbia, Challenger, Discovery,* and *Atlantis.* After the loss of *Challenger* in 1986, a replacement was built and christened *Endeavor.* Upon the loss of *Columbia* in 2003, no replacement was built.

Both orbiter losses were due to the stack design. It is clear from our flight knowledge of the Shuttle era that a manned system with parallel staging only is not really the best idea.

Stack design is basically how the different stages are put together. What we're used to seeing in a serial staged rocket is called the stick design because everything is in a line, like a stick. But the Space Shuttle uses a very different stack design, with the external fuel tank, the solid boosters, and the main liquid engines/crew/payload compartments essentially side by side, hence, the parallel staging descriptive.

In the case of *Challenger,* as everyone knows by now, there was a blow-by at one of the O-rings in the starboard SRB due to low temperatures stiffening the ring the previous night. It welded itself shut, but an unexpected and powerful wind shear broke it open again, and also likely broke the weakened mooring structure at the bottom of the booster. The wind shear not only created an additional thrust vector but also ultimately caused the one booster that was now free at the bottom and still attached at the top to pitch the nose of the booster into the top of the ET. The nose of the booster hitting the tank is what caused the rupture of the ET, which threw oxidizer and fuel out of it. This in turn created an additional thrust vector from the burning hydrogen and oxidizer. The unbalanced forces sheared the stack, tearing the orbiter apart and virtually instantly killing the crew on the middeck, while the flight crew rode the relatively intact flight compartment down to ocean surface impact. Had there been a flight cabin escape system they might have been able to survive. Might. Once again, budget . . . politics.

In the case of *Columbia,* again, the stack configuration was the ultimate problem. The inclusion of an external cryogenic tank required insulation of the tank. This was created using spray-on foam insulation. However, as in the case of the solid fuel of the SRBs, it could not be sprayed on all at once, but had to be done area by area. This created slight discontinuities in the uniformity of the insulation, which made the foam prone to breaking off in certain areas. With the kinds of speeds seen during launch, even a very lightweight piece of debris has enormous momentum and kinetic energy, and so the small fragment (it is believed to have been briefcase-sized) gouged a deep groove in the port wing of the orbiter, punching a 6-to-10 inch (15–25 cm) hole in the leading edge of the wing. Upon reentry, this resulted in overheating of the interior wing structures, leading to sensor failure, hydraulic failure, probable softening of the wing framework, and possible explosion of the pressurized landing gear tires. The end result was that the orbiter disintegrated during reentry at a speed somewhere between Mach 17 and Mach 20, roughly over the state of Texas.

In both cases there was substantial evidence of such potential failures in previous flights, and significant concerns by engineers, scientists, and technicians at, e.g. Thiokol and MSFC for *Challenger*, and the Inter-Center Photo Working Group and the specially-formed Debris Assessment Team (consisting of Boeing, KSC, JSC, and United Space Alliance engineers) for *Columbia*.

If the Shuttle had maintained something approximating the original stick configuration (even if internal—with the crew at the top, *above* all the stuff that goes boom) neither catastrophe might have happened. This should tell us something about how to build future rockets.

Let the scientists and engineers design it and run it, not the damned politicians.

What's the difference between the '60s and recent years for NASA? Why does it seem like they're less innovative, less daring, less "out there" than they used to be?

Because they *are*, unfortunately. But not by choice, as it turns out.

A friend of mine recently had the opportunity to attend an astronaut memorial service for the first time. Since she had a friend, Kalpana Chawla, aboard *Columbia* on its final, fatal voyage, it was rendered all the more poignant for her—and she had been asked to be one of two speakers to provide a NASA historical retrospective for the service. Between them, they covered most of NASA's history, and even overlapped a bit in the middle. And it was he, this other speaker, who revealed to her what made the difference, what caused the change. And she shared it with me.

As it turns out, it was the *Challenger* disaster. It wasn't so much the disaster in and of itself. Astronauts and those who work with them know that spaceflight can be a risky business. Sitting on top of thousands of tons of high explosive, flinging oneself into an environment that is totally inimical to the human life form . . . when you think about it like that, it puts things into perspective. To paraphrase Captain Kirk a little bit, "Risk! Risk is our business! That's what this spacecraft is all about! That's why we're aboard her!"

Actually, we've really done a dang good job keeping those astronauts safe. Anybody who has worked in the space industry in the Rocket City or anywhere else knows that safety is always first and foremost on all of our minds. At any given launch we are all thinking of what we might not have thought of that could come back around and bite us in the butt.

No, the problem that arose from the *Challenger* explosion, the problem that changed the very way NASA operated, was that the disaster occurred with a civilian on board. Christa McCauliffe, who would have been the first teacher in space, was on that Shuttle when it exploded over the Atlantic just off the Florida coast. And the public horror over the event, and the way people related to McCauliffe, caused the fear of a public backlash on Capitol Hill. To prevent that, orders came down from On High that there would be no more risk-taking. All equipment and technology henceforward would be completely proven. Absolutely nothing would be developed especially for the program that in any way constituted an integral structural component or electronic device, without first going through many, many safety checks. Launch window constraints became more rigid. And NASA was forced into a position where it could not even use cutting edge technology available on the street until it was sent through several years' worth of safety testing, by which time it had become obsolete. This added to the cost of the space program. It added a *lot* to the cost of the space program.

This does explain a bunch of things.

The memorial service was particularly bittersweet, my friend said. She had, only a few days earlier, encountered one of the three coroners who had worked the *Columbia* investigation, at a speech and booksigning she gave at her undergraduate alma mater. Seems they had had a chat beforehand, and she got some questions answered that had bothered her since the disaster happened. They weren't the answers she wanted to hear, unfortunately, but at least now she had answers.

When she was asked to give a brief synopsis at the memorial of what happened in *that* catastrophe, she did—but told me later that she couldn't get through it without becoming very emotional. Even though many years have passed since *Columbia* went down, the new information had ripped open the old wound of a friend's loss. According to her, it proved an emotional time for everyone, speakers and audience alike, but it sounds like those who were there more than adequately honored her friend's death, as well as all of the other astronauts who died "with their boots on."

The most disturbing thing to me, however, was that she said the audience consisted of a scant handful. There were probably more organizers there than general public. Yet this event had been widely promoted in the area, to include newspaper and radio—although local television stations apparently chose not to cover it. Public apathy was too great for the general populace to bother remembering those who served us at least as honorably as any soldier, any police officer, any firefighter. In some ways, it was almost as heartbreaking as the disaster itself. Once again, this was more proof to me that the "rock star" status of the American astronaut has long since passed.

In 2011, the Shuttle fleet was retired.[10]

Let's stop for a minute and talk about the Space Shuttle Main Engines (SSMEs), technically known as RS-25 engines. There's a method in my madness, because we'll need to know a little something about them later. The SSMEs are a wonder of engineering. Besides, I'm a rocket scientist after all, somehow I've got to sneak some rocket

---

[10] For more information on the Shuttle, check out:
  http://en.wikipedia.org/wiki/Space_Shuttle_design_process
  http://history.nasa.gov/x15/cover.html
  http://en.wikipedia.org/wiki/STS

science in here either without you, the reader, realizing it—or just by blatantly slapping you in the face with it like I'm doing now.

I have to say this from the start. I absolutely *Love* the Space Shuttle Main Engines. They are an engineering work of art. At first glance, a Space Shuttle Main Engine seems like it should be a relatively straightforward, uncomplicated mechanism—after all, it just mixes fuel and oxygen and lets the two liquids burn, right? Wrong. It's anything but uncomplicated, because the fuel and the oxygen are both stored and introduced into the system in cryogenic (super-cold) liquid states. And in addition to keeping them separate before you're ready to burn them, they not only have to be kept at high pressures, they have to be warmed up before they can be mixed and ignited.

In the case of the SSME, the fuel is liquid hydrogen, commonly known as LH2, and the oxidizer is LOX, liquid oxygen. (These are known as cryogenic fuels because in order to keep them liquid, they must be maintained at extremely low temperatures.) The majority of the exhaust is, therefore, water. Water, and a whole hell of a lot of heat.

The fuel and oxidizer both start in tanks external to the engines. Upon entering the main propulsion system, each encounters its own low-pressure turbopump. The fuel pump is designated the LPHT (Low Pressure Hydrogen Turbopump—axial flow, two-stage, 16,185 rpm), and the oxidizer pump is called the LPOT (Low Pressure Oxygen Turbopump—axial flow, six-stage, 5,150 rpm). These lower pressure pumps enable the high pressure pumps later in the system to operate without cavitation (temporary bubbling in the fluid due to localized changes in pressure) occurring. After passing through the low pressure pumps, the oxidizer and fuel routes differ significantly.

After the LPOT increases the LOX pressure from 100 psia to 422 psia, most of the O2 is then routed to the HPOT (High Pressure Oxygen Turbopump—28,120 rpm) where the pressure is boosted to 4,300 psia. The HPOT itself is made up of two single stage centrifugal pumps, one each for the main pump and preburner pump, on a common shaft driven by a two-stage turbine, which is in turn driven by hot O2 gas from farther down in the system. The flow from the HPOT diverges into three lines. One small line goes back to the LPOT and is used to drive it. The second small line is sent to the oxidizer heat exchanger, where it is heated to vaporization. This O2 gas is then routed all the way back to the oxidizer supply tank to maintain pressure in that tank. The third and largest passes through the main oxidizer valve and into the combustion chamber, but not before a small line branches off just before reaching the valve. This small line feeds to the two preburners, one on the oxidizer side and one on the fuel side, with control valves outside each preburner.

Before we take it any farther, let's get the fuel caught up to the oxidizer.

The LPHT boosts the pressure of the LH2 from 30 psia to 276 psia, again to prevent cavitation in the high pressure pump downflow, and is on the opposite side of the engine structure from the LPOT. The LH2 is then piped directly to the HPHT (High Pressure Hydrogen Turbopump—35,360 rpm). This is a three-stage centrifugal pump driven by a two-stage turbine in the same manner as the HPOT, and boosts the fuel pressure to 6,515 psia. One line emerges from the HPHT and goes through the main fuel valve, then splits into three lines.

One of these is routed through the jacket of the main combustion chamber, where dual purposes are served: 1) the LH2 cools the walls of the combustion chamber, 2) the LH2 is vaporized. The H2 gas is then ducted back to a junction; one path leads back to the LPHT to drive the turbines, the other takes the gas back to the fuel tank to maintain its pressure. After being used for the LPHT, the hydrogen gas is funneled through the hot gas manifold and into the main combustion chamber.

The second fuel line is directed into a spiral tube on the outside of the nozzle itself. This again serves the purpose of cooling the nozzle and heating the hydrogen. It then joins with the third line.

The third line is first routed through another valve, the chamber coolant valve, where it meets up with the prewarmed hydrogen from the nozzle, and travels to the fuel and oxidizer preburners.

Now we have hydrogen gas and oxygen gas meeting up at the preburners, and hydrogen and oxygen in the combustion chamber. It's time for some fireworks.

Each preburner, like the low pressure and high pressure pumps, is located on opposite sides of the combustion chamber. The hydrogen and oxygen are efficiently mixed in the preburners just as fuel injectors would put gasoline into a car engine's cylinder. Dual-redundant spark igniters fire for up to 3 seconds, after which time the fuel in both preburners is ignited and the combustion becomes self-sustaining. The exhaust pressure from these preburners drives both the HPOT and the HPHT. It is then released into the hot gas manifold and thence into the main combustion chamber.

Meanwhile, a similar set of "spark plugs" ignites in the main combustion chamber during the engine start sequence, likewise creating self-sustaining combustion in about three seconds. At this point, the engine is fully operational and wide open.

The Shuttle actually uses the SSMEs to do some steering by doing what is called thrust vector control. The SSMEs are on gimbals that point the nozzles to steer the spacecraft.

The LPOT and the LPHT are mounted to the aft fuselage structure. The rest of the engine assembly is what is mounted on the gimbals, with the lines between the low pressure pumps and the rest of the assembly consisting of flexible bellows. This enables the thrust to be vectored (angled in a particular direction).

Increasing and decreasing the thrust all boils down to two little valves: the oxidizer and fuel preburner oxygen valves. Adjustment of these valves in turn controls the chamber pressure in the preburners, which regulates the speed of the HPOT and the HPHT. That regulates the rate of flow of O2 and H2 into the main combustion chamber, which determines the thrust. The two valves operate together to maintain a preferred 6:1 propellant mixture ratio. This gives us the lovely "blue diamond" exhaust pattern so beloved by rocket and jet enthusiasts.

"But what about the main valves?" you ask. Frankly, if the engine is running, the main valves for fuel and oxidizer are wide open and stay that way. Remember that chamber coolant valve in one of the fuel lines? That regulates how much hydrogen goes through the nozzle cooling tubes. It's wide open too, before launch, and as long as the throttle setting is 100 to 109 percent, it stays that way. When the engines are throttled back (they can throttle down to 65 percent), it can afford to be closed a bit, too, down to no more than 66.4 percent open. All five valves are hydraulically actuated and electrically controlled. In the event of an emergency shutdown, the main propulsion system (MPS, the system that controls all three Shuttle engines) can use its helium pneumatic system to close all five valves, as well as aid in a propellant/oxidizer dump in the event of a launch abort of any kind.

I didn't go through this to hurt your noggin or to show off how much I know about the SSMEs.[11] The reason for adding this last couple of pages of highly technical details of how the SSMEs actually function is to show just how complex a rocket engine truly is. With something so outrageously complex never should there be a politician in the design loop. Now there is a time and place for everything. The politicians are needed to help keep the budgets put in place to build the rockets, but they should not be making design choices or flight choices (like flying in freezing cold weather if it is outside the designed operational specs). If rocket science was easy politicians would do it.

So if we no longer have the Saturn V or the Space Shuttle, what other options do we have? NASA had just proven our best and quickest chance of exceeding LEO when it successfully tested the Ares I rocket in 2009. Immediately following that successful test, the program was canceled—for political reasons, once more. Where does that leave us?

NASA is currently trying to recover from the cancellation of the Constellation/Ares program by working on the Space Launch System, or SLS. It is intended to be a heavy lift system that is "safe, affordable and sustainable,"[12] and to carry the Orion capsule built for NASA by Lockheed Martin for the Constellation program and able to return to the Moon or go to Mars. The SLS will also provide transportation to the International Space Station and commercial launch capability. It is being developed and managed out of the Marshall Space Flight Center here in Huntsville, Alabama. But from the start, I feel obliged to point out that SLS was politically chosen based on heavy lobbying from various commercial rocket organizations like United Launch Alliance, Lockheed Martin, Boeing, and SpaceX.

---

[11] More information on the SSME:
   http://science.ksc.nasa.gov/shuttle/technology/sts-newsref/sts-mps.html#sts-mps-ssme
   http://en.wikipedia.org/wiki/File:Ssme_schematic.svg

[12] http://www.nasa.gov/exploration/systems/sls/sls1.html

The intent is to utilize as much already developed technology as possible, resulting in decreased cost and hopefully increased safety. This includes Shuttle designs and Constellation as well. Don't worry, for now I guess, about the fact that we were five years into the design and test of Constellation when we decided to throw all that out and start over. Again. Fortunately, if we stay with this plan, we might end up with a decent rocket, as the SLS is very similar in design to the previous Constellation Program's Ares V.

The first stage, or core, uses the RS-25 engine built by Pratt & Whitney Rocketdyne (a fully owned subsidiary of United Technologies), which is more commonly known as the Space Shuttle Main Engine (SSME), like the SSME, it will be LH2/LOX fueled, with the main part of the core consisting of a modified Space Shuttle External Tank (ET), manufactured by Lockheed Martin at NASA's Michoud Assembly Facility in Louisiana. The Block I (early) design calls for 4 SSMEs, while the Block II design calls for five. In addition, Block I will have two modified Shuttle Solid Rocket Boosters (SRBs) strapped alongside the first stage, made once more by Thiokol Corporation in Utah. These SRBs will have either four or five segments, depending upon thrust need. Block II boosters are currently out on bid. They may be solid or liquid, depending upon the winning design.

The combined thrust of the Block I first stage and the two SRBs is on the order of 8.87 million pounds (39.47 million N). The Block II first stage will have a thrust of 2.1 million pounds (9.3 million N). The total initial thrust can't be determined until the strap-on booster contract is awarded and the design finalized.

A domestic version of the Russian NK-33 engine (originally intended for the Soviet N-1 lunar booster, which booster was never successfully completed) is being developed by government contractors Aerojet and Teledyne Brown, under Aerojet's license to Russia, to compete with the Space Shuttle Main Engine as the engine of choice for the SLS first stage. Each individual engine would have a thrust of 500,000 pounds (2.2 million N). This engine uses LOX and RP-1 as fuel.

But once again, we would be depending on the Russians for our rockets, when we have a perfectly viable American-made engine in the SSME. A damned great American-made engine in the SSME. Moreover, the specific impulse (thrust per amount of propellant burned per unit time) of the NK-33 is only 331 seconds of burn time, with an exhaust velocity of 3,240 m/s, whereas the SSME has a specific impulse of 453 seconds, and an exhaust velocity of 4,423 m/s. Clearly the SSME is the more efficient, more powerful engine, in addition to needing no further development. Also, flight history and knowledge of the SSME makes it safer to use than a new, untested engine that we have to buy from the Russians.

The second stage of the SLS is essentially a modified Delta stage (called the interim Cryogenic Propulsion Stage or iCPS. The stage is built by Boeing and Lockheed Martin under the United Launch Alliance label, whereas the engines are built by Pratt & Whitney Rocketdyne) in Block I, using a single RL10B-2 engine, which engines were also used in the Saturn I second stage, and somehow were not lost in the interim—most likely due to being used on other rockets like the Delta rockets. It has a thrust of 24,800 pounds (110,100 N) and uses LH2/LOX fuel. Block II, which will be Earth-departure capable, will use three J2-X engines built by Pratt & Whitney Rocketdyne, a variant on the J-2 used in two of the Saturn V stages that were being developed for the Ares I in the Constellation Program. They are capable of being re-started after shutdown, which makes them ideal for a trans-lunar injection burn or an Earth escape burn. It also burns LH2/LOX, and has a thrust of 880,000 lbs (3.93 million N).

Overall, Block I will have a payload capacity of 154,000 lb (70,000 kg) to low Earth orbit—the rough equivalent of 40 SUVs. Block II is anticipated to have a payload lift capability of 286,000 lb (129,000 kg) to low Earth orbit—75 SUVs' worth. To lunar orbit, Block I can carry 145,200 lb (66,000 kg), and Block II up to 286,000 lb (130,000 kg). Compare this to the Saturn V, which could carry 262,000 lb (119,000 kg) to LEO and 100,000 lb (45,000 kg) into lunar orbit.

The NASA presentation to the Senate in 2011 proposed an $18 billion budget through 2017 and initial, unmanned launch. Unfortunately most government programs such as NASA, chronically starved, have learned to underbid just to get the program funded, then hope for "plus ups," budgetary additions to support the real program costs. Unofficial estimates of the final costs are $41 billion for the first four Block I launches, with a Block II vehicle not available until 2030.

What NASA should be doing and Americans should be demanding is that the budget be $100 billion per year for the next decade. We just blew 30 times that on crazy bailouts that we have nothing to show for. In the space program's case we'd create millions of jobs and thousands of spin-off technologies and we'd create a new generation of highly trained technology professionals within this country.

Let's compare the proposed SLS costs to the Saturn/Apollo program. The total cost for the Saturn V rocket development from 1964 to 1973 was $6.5 billion *in the currency of that time*. Most estimates of inflation on that figure would place the modern dollar value at approximately $45–65 billion. Not so very different when you view it from that perspective.

CHAPTER 7:
COMMERCIAL SPACE AIN' T
SO COMMERCIAL

You might think that our Junkyard Iron Man suit was the most expensive build we did, but it wasn't. We used mostly junk parts, a few new motors, some fairly inexpensive servo controller circuits that we bought online, and a bunch of fiberglass. The biggest cost on Iron Man was that it took longer to build than expected so we had to pay camera crews for an extra weekend. The same thing went with the submarine. It wasn't that expensive. Maybe that's why we sank it?

The most expensive project we did was the double barreled rocket. Rockets are expensive. Actually, the motors matched the cost of the Iron Man suit alone. We used 2 N5800 solid rocket motors. We spent another thousand bucks or so on a parachute that we never got to deploy. And then the material for the rocket body and thrust structure and the large man-capable capsule was another couple thousand bucks. Somewhere along the way I guesstimated that we spent around $10,000 to build that rocket, or would have if we hadn't been able to find some of the stuff in salvage yards.

So, just to go a thousand feet high or so with a man-sized payload (had the rocket not failed) cost us about $10K. To send a guy 300 times higher to reach the edge of space should be much more expensive. If the cost is linear with altitude then it should cost about $3,000,000! Three million dollars! And that isn't even to orbit. That would be a quick fifteen-minute flight like Alan Shepard's first suborbital flight. To go orbital has to be at least ten times more expensive. That sort of adds up as the Russians charge about $25 million to put a single tourist into orbit and on the ISS.

I loaded the igniters into the large solid rocket motors before we raised the rocket on the tower. It was easier that way, but a whole lot scarier!

So for us Rocket City Rednecks, right now anyway, we simply don't have the funds to build a man-capable rocket that could reach space. This type of flight experiment is in the realm of multimillionaires and billionaires, corporate entities, or governments. That ain't to say that we haven't been thinking on how to do it! Somehow soon, I plan to figure out how to build a small-payload orbital rocket. I've even spent some time talking with my buddy Tim Pickens, who was one of the chief builders of the hybrid rocket for SpaceShipOne. We have some ideas, but they do require a budget bigger than we can muster for a single episode of the show.

We have flown a man-capable rocket to higher altitudes, but I'd hate to give away spoilers from Season Two of the show. Stay tuned and you'll see that with a similar budget a group of guys can build a heck of a rocket in a weekend and have a pretty amazing flight.

Just think if we had real budget to play with! Burt Rutan showed

Once the man-capable-sized rocket was erected somebody had to climb up there and set up all the flight computers and explosive charges.

us all what can be done with some innovation, hard-working can-do folks, and a few tens of millions of dollars. Hopefully, if not us, some smart kids will get out there and figure out a way to build rockets cheaper that can reach space. We just need to keep pushing that atmospheric boundary with what few dollars and resources we can scrounge. Sooner or later, hopefully sooner, we'll figure out a way to get there in a manner we can all afford. Well, that is, all of us who want to get there. I bet you couldn't pay Rog any amount of money to ride on a rocket. Me? Put me in, coach. I'm ready to play.

The Double Barrel Rocket was a sight to see on the pad. Like all rocket experiments, in some form or fashion they end with a blast. This one was certainly no exception.

★ ★ ★

Fine, you say. But space should be commercialized! We should let the companies in the business of making rockets worry about that. Then we can save the money in NASA's budget for other things. Or maybe we don't need NASA at all. Heck, then we could kill NASA!

So let's talk about commercial space exploration. The commercial space enthusiasts would have everybody believe that is an industry ripe for picking, with billions and billions of dollars waiting to be had. The commercial space enthusiasts have us convinced that all we need are some slightly easier government regulations and some big prizes, like $20 million to be the first person to put an astronaut into orbit and bring them home safely and do it again a couple weeks later with the same vehicle. Or, like the Google lunar prize to land a rover on the Moon for $20 million.

Spaceship One was a very successful mission. It was very dangerous. It was a stunt. It was based on a novel approach. And it was cool as hell. On the other hand, if it had made flying Spaceship One to suborbital altitudes a lucrative business model, Spaceship One would still be flying to suborbital every day, or at least every few weeks. The reward, the so-called X-prize, that Spaceship One allowed Paul Allen's team to win was $20 million. Paul Allen spent at least that much of his own money to build the thing. And that's not including all of the money and time and effort that Scaled Composites put into it, Tim Pickens' team at Orion Propulsion in Huntsville, Alabama, and all the other smaller subcontractors or individuals who donated lots of their time to this effort with hopes of winning the X-Prize. It didn't foster in a whole new era of commercial flight to space—at least not yet. And to offer $20 million to put a lander on the moon? This is where the commercial space industry is relying upon the poor math education of the general public—or at least they are relying on the hopes that nobody in the general public will do the math correctly.

And then we have the billionaires starting companies with the business model that other billionaires will want to ride into space and vacation there paying large sums of money to do so. The Russians have been flying billionaires to Mir and the International Space Station for about $20–$25 million a pop. That's just to go into low Earth orbit, and they have only done this a few times. The only way that they made money off these trips is the fact that they were going to the International Space Station anyway, to carry supplies and other astronauts. They had extra space in the Soyuz space capsule so they sold it for $25 million.

A good rule of thumb to go by if you have a satellite that needs to be launched into low Earth orbit is somewhere between ten and twenty thousand dollars per pound to get up there. So think about an average human astronaut. The astronaut weighs about two hundred pounds fully clothed in his jumpsuit but with no other gear. So that means it costs somewhere between $2 to $4 million just to put the astronaut with no life support systems or personal belongings into low Earth orbit. And that is just to put him into space. Bringing him home safely costs at least as much, if not more. Now these numbers, the $10,000 to $20,000 per pound, are numbers generated from years of NASA and the Air Force and the commercial industry placing satellites into low Earth orbit. And these numbers are based on unmanned launch vehicles.

Unmanned launch vehicles do not have to go through the immense amount of inspection and safety verification that a manned vehicle does. To man-rate something for spaceflight requires quality testing of all of the components in your spacecraft. The process increases the cost by at least a factor of ten. This puts safely flying an astronaut to low Earth orbit at somewhere between $20–$40 million and roughly the same amount again to bring him home safely. So just to fly an astronaut safely into low Earth orbit to the space station and home again should cost somewhere between $40–$80 million. Once again the Russians were able to sell rides at $25 million a pop because they were already going and they had an empty seat. The way their rockets were, they would've had to fill the capsule with some sort of ballast materials anyway; why not a paying customer? Actually, a paying customer was more likely the United States of America paying the Russians to carry supplies to the space station, which in essence supplemented their commercial space venture—that America made *no* money on.

There are also a few enthusiasts' companies like Bigelow, Space Adventures, Virgin Galactic, etc. These guys want to build hotels in low Earth orbit, fly people to the Moon, and fly people regularly into space. They are all convincing us that there is a business model to support these efforts. It's been reported that the trips to the Moon would be in a modified Russian spacecraft, and they would do a fly-around and come home, not making a lunar landing. The seats on the vehicle would go for something like $100 million apiece. Therefore, this business model requires there to be people able to afford $100 million for the trip. The vehicle is a Russian one so you have a space-craft that will hold one captain who works for the company and two paying astronauts. This suggests that the Russians believe $200 million per flight is enough of a business model. How does this help the rest of us with the space plan?

Let's assume that there are one million billionaires in the world. (According to *Forbes* Magazine, there are actually only 1,210. If there are ten percent of those billionaires who would like to go into space and can afford $100 million per seat, that means there are one hundred thousand potential customers. The number is probably more like one percent, which would mean there are ten thousand potential customers. Of those ten thousand probably one percent of them can meet the physical requirements for the trip. That means about one hundred actual potential customers. That results in a possibility of fifty flights. Well, that's a viable business model for the near-term assuming that the Russians could put together fifty flights to the Moon with fifty to one hundred paying customers. If you recall, Dennis Tito, on his return from space, said that he would definitely not go back again. It's likely that a trip around the Moon would be much more startling, physically demanding, emotionally and mentally demanding, and just flat out scary. I would be willing to bet that potential one hundred customers would dwindle dramatically after the first couple of missions. And if there were ever an Apollo 13 type incident, or even worse where there were injuries or deaths, that would likely kill that business model.

This does nothing for humanity's exploration of space. Oh sure, if a few playboy billionaires can go and do it, eventually it may be able to help to reduce the cost of getting more people into space, but that's unlikely in the near future.

And then there are those who would convince Congress that the best way to go to space is to let the industry develop the vehicles and then pay for rides in the spacecraft. United Launch Alliance has lobbied deeply to create this attitude. The Saturn V and the space shuttle both had one to two million moving parts or at least parts of some significance on each of 'em. It's likely that each of these parts cost somewhere between $10,000–$200,000 each just to make *one*. That puts the cost of the shuttle or the Saturn V somewhere between $10–$100 billion, as we discussed in greater detail in the last chapter. The estimated cost of an expendable launch vehicle like the Delta-4 heavy or the Atlas-5 is around $200 million. These expendable launch vehicles are not man-rated and they do not have reentry and landing systems built into them. Assuming at least twice the cost, and that the expendable launch vehicles with a reentry system and landing system could be built for a half a billion or so, this sort of gives us an idea of the difference between flying an unmanned rocket and a manned rocket. NASA authorization legislation in 2000 capped the cost of a Shuttle flight at $380 million. A Titan 4B, with approximately the same payload capabilities, costs $400–$500 million per launch, and an EELV Heavy would be roughly $232 million. However, a single Saturn V launch, per the *Encyclopedia Astronautica*, cost $431 million *in 1967 dollars!* That's about $2.5 billion today.[13]

So United Launch Alliance and SpaceX have convinced the Obama administration and the current Congress that they should be able to build our manned launch vehicles and just let the government catch a ride to wherever it is they're going to go. Unfortunately there is not enough of a business case for this approach. What this actually does is create a new industry that must be government subsidized or it will fail just like the airline system in the United States. If our government would get out of the way and quit helping out the airlines and let them succeed or fail they might become a viable business. However, the airlines are such a part of our national infrastructure and methodology for constructing new business that the government feels it must be involved to maintain its operation. Perhaps this is correct; perhaps it's incorrect, but the airline business is far less complex than sending man to the Moon, Mars, or beyond. The airlines could likely stand up by themselves now if we force them to. But the space industry is much more immature than that.

So instead of creating a whole new industry that relies on government programs to keep it alive under the false pretenses of it

---

[13] http://www.thespacereview.com/article/146/1

being a commercially viable industry, it is more likely that an effective space program with a central goal and nexus, meaning NASA, makes more sense, at least in the near-term. Nobody is stopping SpaceX and ULA and Virgin Galactic from building their own spaceships and flying into space—or at least we shouldn't be—but that is not going to keep Americans in front of the international space race and atop the new discoveries in science and engineering that will keep us a thriving superpower in the future. What would be better than to supplement ULA and SpaceX and the other commercial space industries would be for us to have a national space program as their main business and the others as venture capital investments.

The Falcon launch vehicles were designed and built by the SpaceX Corporation. For our purposes, there are two of principal interest: the Falcon-9 and the Falcon Heavy.

The Falcon-9 is a two-stage-to-orbit launch vehicle that uses RP-1 and LOX. It has already been successfully launched several times, including at least one launch with the Dragon spacecraft atop the stack. The stack has an overall height of 178 feet (54.3 m), diameter of 12.0 feet (3.66m), and a mass of 735,000 pounds (333,400 kg). It is considered a medium-lift launch vehicle, capable of lifting 23,000-58,700 pounds (10,450-26,610 kg) to low Earth orbit, or 9,800-33,100 pounds (4,450-15,010 kg) to geostationary orbit. Launch site options include Cape Canaveral (KSC/USAF), Vandenberg AFB, and Omelek Island in the Kwajalein Atoll's Ballistic Missile Defense Test Site.

The first stage uses SpaceX's own Merlin-1C engines, nine of them, to be specific. With a thrust of 125,000 pounds (556 kN) per engine, this yields a total first stage thrust of about 1.1 million pounds (5 million N). The burn time is 170 seconds (2 min, 50 s).

The second stage uses a single Merlin-1C engine, with a burn time of 345 seconds (5 min, 45 s) to achieve orbit. SpaceX is pushing the development of the Falcon-9 and Heavy in hopes of becoming a true contender in the large launch vehicle industry in the near future. Time will tell how well they do.

The Merlin engines incorporate legacy technology, such as the type of fuel injector system used on the Apollo-era LM engine, but eliminates a separate hydraulic system for vectoring the thrust, using the high pressure fuel itself for the purpose. This in turn eliminates one potential failure mode for the engine.

Both stages are intended to be recovered and reused, and to that end have been outfitted with parachutes and ablative cork exterior layering, as well as protection against a corrosive saltwater environment. However, to date, no stages have been recovered that I know of.

In 2008, NASA contracted with SpaceX for twelve supply flights to the ISS, for $1.6 billion. The contract has an extension option for additional flights totaling up to $3.1 billion.[14]

According to SpaceX, the Falcon Heavy is—or will be since it has not yet flown—the world's most powerful rocket. It has a payload capacity twice that of the Space Shuttle, at 117,000 pounds (53,000 kg) to low Earth orbit, 42,000 pounds (19,000 kg) to geostationary, 35,000 pounds (16,000 kg) into trans-lunar injection, and 31,000 pounds (14,000 kg) into a Martian trajectory.

It is essentially a modified Falcon-9, with two additional first stages strapped to the sides of a conventional Falcon-9. This makes for a total of twenty-seven Merlin engines and 3.8 million lb (16.8 million N) of thrust. This is the equivalent of fifteen wide open, full throttle Boeing 747 jet aircraft. It uses triple-redundant avionics and, like the Falcon-9 of which it is composed, is capable of sustaining several engine failures without mission failure. Also, each engine is encapsulated with a special shell that prevents crossovers (fire, debris, etc.) from affecting the other engines. Actually, the redundancy of the Falcon system—the fact that it has many engines on the first stage—is one of its safety points that I like. A couple engines could fail and you'd still have several firing away to enable you to limp onward to some abort scenario.

The stack is 227 feet (69.2 m) tall. It is still only 12 feet (3.66 m) deep, but because of the three-abreast configuration, is some 36 feet (11 m) wide. It has a total weight of 3.1 million pounds (1.4 million kg).

NASA's Ames Research Center is currently developing a Mars mission plan that incorporates a Falcon Heavy and Dragon capsule, known as Red Dragon. If the proposal is accepted and funded, it may

---

[14] http://www.spacex.com/falcon9.php
[http://en.wikipedia.org/wiki/Falcon_9

be slated for a launch as early as 2018. But, I definitely have to say here that there is no political backing for a Mars mission as of yet so there is no budget for it. Also, NASA can't just give SpaceX a contract to go forth and do this. They would have to compete for the mission with other spacecraft contractors in an open acquisition process.[15]

I think that one of the biggest problems between government space plans and commercial space plans over the last fifty years has been the fact that the two somehow couldn't coincide with each other. Burt Rutan is often seen giving lectures on how NASA is keeping Americans out of space. I've watched him give this speech before. I really don't believe that government and commercial plans have to be at odds with each other. In fact, I believe they can complement each other. And I'm not so sure that Rutan really believes that either. But if he does, I would like to understand more about why he does. Rutan is a smart guy.

"Son," Grandma would say. "I know you're a smart boy. But no matter how smart you are somebody out there always has a better idea. And, if y'all was smart you'd listen to 'em." Grandma was right. We should listen to everybody who has a potentially good idea and/or counterargument and go from there.

Rutan is often telling us that it is NASA and the federal government keeping us out of space based on regulations and budgets. Now, there might be some laws and some FAA types of pressures and nowadays some pressures from the ATF regarding rocket fuel and explosives that may make it difficult for commercial entities to develop spacecraft technologies. However, that may just be the nature of the beast, especially in the day and age that we live.

It's hard to see just exactly where the Space Shuttle injured Paul Allen and Rutan's Spaceship One effort. It is likely that most of that is political lip service. So, if a person were to decide they were going to build a launch vehicle and gain access to space, I would guess they could use the Spaceship One litmus test to show that it can be done through all technical, legal, and political avenues. There's no reason to believe that you can't have both the Space Shuttle and the

---

[15] http://www.spacex.com/falcon_heavy.php
   http://en.wikipedia.org/wiki/Falcon_Heavy

Spaceship One, or another space launch system, whatever, maybe government funded and controlled, and run as well as a completely private entity like Paul Allen and Scaled Composites and Burt Rutan's Spaceship One. And in fact, government incentives for industries to develop completely commercial access to space is a really good idea simply because there isn't a real business case that is supported yet.

One thing to learn from Spaceship One and from all the past NASA programs is that innovation can come from either a government space program or a private or commercial space program and it is not necessary for them to be inclusive of each other, coupled to each other, or even derived from each other. However, they certainly could benefit from each other. The Spaceship One reentry concept was nothing short of brilliant and should really be investigated for future reusable launch capabilities. If y'all was smart you'd listen to my grandma and consider a reentry system like that of Spaceship One.

As far as commercial vendors creating launch vehicles and access to space, from which the government is likely their only customer anyway, that is very suspect to me as being economically viable. The Falcon-9, the Atlas-5, and the Delta-4 are all very good launch vehicle concepts. The Falcon-9 will soon be online and is likely to be competitive to the United Launch Alliance evolved expendable launch vehicle (EELV) technologies. But to believe that either of these concepts were purely commercially derived is very naïve.

The Delta and Atlas rockets were derived from competition between Boeing and Lockheed Martin run by the Air Force and NASA, called the EELV program. And DARPA paid for most of the development of the original SpaceX Falcon launch vehicle systems. This is not a commercially developed vehicle because it wasn't from private investor funds. In fact, it was from government funds in a government competition and therefore is not in any definition of the word commercial. So, it's bordering on ridiculous at this point to just jump up and assume commercial vendors for future government space access is a good idea.

In order to develop any new launch vehicles, it would require an immense amount of funding and it's unlikely that the commercial competitors will want to spend their internal research and development and marketing dollars to develop the systems when there is only the likelihood of a scant few products that can be purchased by the government. There might be a handful of other private sources buying them, though that is unlikely. There just is not a large number of sales to be had so the few products produced would be very expensive. There is no demand for the product to make this a viable commercial business, at least none that I can see right now. That is not to say that some smart guy won't figure it out in the future. I hope he or she does.

Let's determine just how much of these launch vehicles are truly commercial, in that the research, development, and testing were funded entirely by the company or companies or investors involved in building it. The original Atlas rocket was an intercontinental ballistic missile (ICBM), commissioned from the Convair division of General Dynamics. (Convair was eventually absorbed by Lockheed-Martin, which is part of the ULA consortium.) The initial contract for development of prototypes was awarded in 1946 for $1.89 million, by the U.S. Air Corps (it wasn't even the Air Force yet). By the time they had a successful, well-developed ICBM, the Atlas project had cost the government some $8 billion—in 1950s dollars. Very quickly NASA adopted it, adding stages to develop the Atlas-Agena, upon which Mercury missions were launched. The modern Atlas-5 is developed from technology used in the earlier Atlas-3 (continued development from NASA and the USAF—an additional $397 million, division between the company and government uncertain) and the Titan, which was developed for the USAF at a cost of $432 million. Atlas-5 is favored due to its use of proven technology.

Let's say that the government funded $8.5 billion of Atlas-5 development in pre-inflation dollars. They're launched from a government facility—the Kennedy Space Center—with civil service and paid government contractor staffing, and each rocket is paid for by the government, at a rate of $147-187 million per launch. NASA just contracted with ULA to launch a Mars probe for $187 million. ULA just signed a DoD contract worth $1.516 billion for five Atlas-5 launches and four Delta-4 launches. And NASA is throwing $6.7 million into the pot for developing an emergency system test bed, needed to man-rate the Atlas-5. ULA is only putting in $1.3 million.

What about SpaceX and the Falcon Heavy? Well, listen to this quote from the company president, Gwinne Shotwell:

"Based on SpaceX's proven track record in scaling tenfold in thrust from Falcon 1 to Falcon 9, we are confident we can scale tenfold again and develop a heavy-lift launch vehicle with a 150 metric ton to orbit capability. We can do so for no more than $2.5 billion, within five years, on a firm, fixed price basis with payment made only on achieving hardware milestones."

In other words, give us two and a half billion dollars, we'll deliver your heavy-lift launch vehicle. And CEO Elon Musk guarantees the cost per flight will be no more than $300 million. So the government pays for the development, then has to buy the individual rockets. Clearly this is not true commercial development as the government would pay for the development and then again pay for the product.

We've simply been talking about access to space and how silly that sounds from a commercial business standpoint, the reason being, not many likely sales. When you think about space transportation and orbit-to-surface access at the Moon or Mars it becomes even sillier. With only a few nations likely to plan and fund it, it's very difficult to see a commercially viable business based on building a lunar surface access module. There just aren't any other uses for them other than manned missions to the Moon. And until the world is filled with trillionaires, economically viable commercial concepts for manned access to the Moon are just beyond any reason whatsoever.

CHAPTER 8:
FORTUNE FAVORS THE BOLD!

I've been thinking about using lunar craters as radio telescope dishes for about ten years. I've read about craters and Moon dirt and have built radio telescopes before. In fact, I built one in high school for a science fair project. But I used giant Yagi antennas instead of dish antennas. Yagi antennas are the ones you used to use to pick up local television channels. The antenna looks like a central rod that you point towards the TV station and there are other metal rods attached to the central one at ninety degrees. It looks sort of like a fish skeleton.

At any rate, I have always wanted to go to Arecibo in Puerto Rico and look at the really large dish there built in a sink hole. That one, though, is actually a shiny, reflective metal structure like the dish you use on your house to get digital satellite television signals. No, what I really wanted to do was to build a dish using just dirt. But, I never followed through on it other than writing a few proposals to NASA and DARPA about it over the years.

One of our first season ideas was to build a radio telescope and listen for alien signals. The producers wanted to know if we could build a very large array out of junk stuff. I told them we could, using a bunch of digital satellite dishes. So we set out and found a few and hooked them all up. They worked well. We built what is known as an interferometer using three different satellite dishes. We picked up signals from the Sun at around 10 GHz but never heard a peep out of E.T.

While we were doing the episode, I mentioned to the director and producer about how cool it would be if we just dug a hole in the ground and made a big dish like at Arecibo. Immediately they wanted to hear more and were curious if we could pull that off with the stuff we had. I told them sure we could. This was before I told Dr. Pete to go off and do some modeling of the antenna on his computer to see how high our tower would have to be for the feedhorn.

Go look at a satellite dish. There is a little electronic doohickey called a feedhorn held up by a rod or pole or tripod over the center of the dish. Think of it like this. The dish is just a magnifying glass that focuses light (in this case radio waves) from space to a small point. We place an electronic receiver/amplifier at the focus and can pick up those signals. The bigger the dish the more signal we can get. In other words, the bigger the dish the weaker, fainter, further-away signal we can receive.

One problem is that the dish and the pole holding the feedhorn in place scale up together. So, a sixty-foot-diameter dish would need a pole to hold a feedhorn in place somewhere like twenty or thirty feet above it. When Pete came back to me a good half hour later and showed me his calculations my first thoughts were that we would have to build a scaffold and that would take too long. But then he showed me a "complex telescope" design that had a second reflector that (on the computer) floated about twelve foot above the main or primary reflector dish and brought the signal back down to the center of the dish where we would put the feedhorns.

Pete's first analysis though had the feedhorns still seven or eight feet high which meant we'd still need some pretty high tables. Then I recalled how Arecibo worked. The secondary reflector is upside down and actually spreads the beam out a little, making the focus slightly further away. Once Pete turned the reflector over in his calculations the beam came to focus right on the ground. Aha! Now all we had to do was find a big dish, turn it dish facing up, and find a way to keep it floating in the air.

Well, that's where Daddy came in. He had a buddy with a huge dish and needed it out of his yard anyway. The dish hadn't been used for years. We got it and Pete and I did some quick calculations as to where to put it and how big and deep our main hole needed to be. Then it was Michael's turn to find us an excavator and get us a hole dug.

Once the hole was dug and the secondary dish put in place with ropes, poles, and 2x4s, we wired it all together. And almost immediately we got signals from the Crab Nebula. It worked perfectly. One time we did get a weird repeating signal for about ninety seconds and then it went away. We didn't see any airplanes anywhere. I checked, there was a satellite that passed over us about that time. It is my guess that we picked up telemetry signals from a low Earth orbiting satellite. Rog likes to think it was a signal from E.T. I doubt that.

But what we showed here in one weekend was that we could use a crater as a dish antenna. In fact, I did some more research and wrote it up as my second Ph.D. dissertation. How many shows out there can say they spun off an actual Ph.D.? And, it sets the stage for moving forward and doing experiments on the Moon using the same concept. There are plenty of craters there that are very, very large. We could likely have radio telescopes so large that we would be able to learn many new things related to cosmology and the nature of the universe. The telescopes we could put together on the Moon using this concept would dwarf anything we could do on Earth. We need to look into this. But in order to use craters on the Moon we'd have to go back there first!

We must go back to the Moon.

★ ★ ★

Why build a lunar base? Even a lunar colony? It's really pretty simple: Practice. Not to mention resources.

The Moon is close to home so it's safer to get to and from. Doesn't mean it's safer than going to Mars or to an asteroid or to other planets that are farther away, but the trip is shorter—so malfunctions and accidents and the like are less likely to happen—and we might be able to stage a rescue if something *did* go wrong. Sort of like Project Gemini—do all your learning close to home where somebody can come get you if you don't do something right.

The Moon is also very interesting in that we believe there should be water ice, frozen in some of the dark, shadowy craters on the Moon. Most of the minerals that are on Earth should be on the Moon, too. We detected radon gas in almost all missions orbiting the Moon, which suggests that there is uranium on the Moon. You see, radon gas is a byproduct of uranium decay. Based on analysis of samples brought back by Project Apollo, the lunar soil also has a bunch of metals in it like titanium, magnesium, aluminum and iron—metals important to our technologies because they're strong and lightweight—as well as lots of oxygen. *And* the act of extracting the metals provides air for the extraction crew to breathe.

So there's all the resources on the Moon that would be needed to build an infrastructure and possibly a habitat and community. I certainly believe that our next imperative should be to build a colony on the Moon. We should learn how to sustain that colony indefinitely. And, we should learn how to make that colony self-sufficient, prosperous, and growing like any Earth-based city.

Now I find it interesting that a former governor from Massachusetts, while he was running for president, made a statement that if one of his vice presidents came to a company meeting saying they wanted a couple hundred billion dollars to build a colony on the Moon, he would fire him. This just goes to show how little Governor Romney understood about space exploration and the spinoff technologies and industries created by it. The governor should have put his staff to work understanding and developing a better policy about the Moon. In fact a lot of people don't realize just how little investment was made in the space program and how many spinoff technologies and industries were created by it. To name some:

- ❈ the instant replay,
- ❈ MRI and CAT scan imaging technologies,
- ❈ lightweight materials that are used on everything
  from baseball bats to bicycles to artificial joints
  to prosthetic limbs,
- ❈ laser ranging and detection technologies
  that were invented right here in Huntsville
   for the docking procedures—
  people use those every day on the
  golf course to see how far away the green is,
- ❈ food purification concepts,
- ❈ "instant" freeze-dried food,
- ❈ sensor technologies,
- ❈ Velcro applications (Velcro was actually invented before
  but the space program really latched onto using it),

❀ Mylar,

❀ memory foam,

❀ the beginnings of electronic miniaturization and ruggedization,

❀ solar cells,

❀ infrared thermometers,

❀ anti-icing systems,

❀ video enhancement and analysis,

❀ fire-resistant materials,

❀ lighter, stronger firefighter gear,

❀ portable cordless vacuums,

❀ water purification,

❀ embedded web technology in houses,

❀ powdered lubricants,

❀ radiation protection,

❀ radiation dosimetry equipment,

❀ structural analysis.

And the list goes on and on.

Each of these technologies put countless people to work with good paying, high technology, high skillset jobs. Millions of jobs have been created by going to the Moon just a handful of times. Millions of jobs!

If I was a shareholder and member of the board, and Governor Romney fired my vice president who had the idea for this investment, my response is simple and clear.

I would fire Governor Romney.

And I'm not saying this because I emotionally want to see a space program succeed. Oh, I definitely do, and the emotions are strong. But the numbers add up and show that it is a major, *significant* investment in technologies, in improving the education of our next generations, increasing American morale, and stimulating massive numbers of growth industries that are derived from spinoff technologies. Governor Romney, you should do your homework. In fact, *all* you politicians should do your homework. The space program should be more than just a soundbite for political use. It is an investment strategy for America's future in high technology, and competition for having high quality jobs in our future.

Now I want to point out here that this chapter and the entire book actually makes no claim of support to any political party or candidates. In fact, I think what we are showing is that, when it comes to space, they're all boneheads. We have in our laps the easiest way to bring America's greatness, economic future, technological superiority, and general morale back. A broad reaching, bold, and exciting American Space Plan would do this. It's time to give our astronauts back their "rock star" status. You politicians out there need to hear what I'm saying. If y'all was smart you'd all jump onboard with a bold American Space Plan. *Fortune favors the bold.*

CHAPTER 9:
BEFORE GOING TO MARS
WE NEED TO PRACTICE

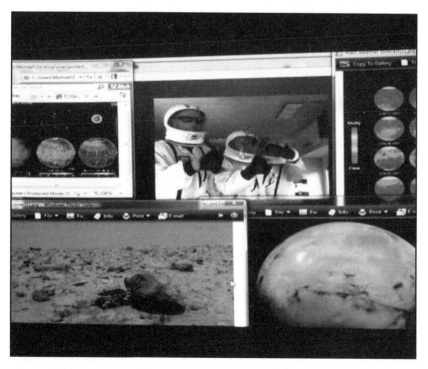

The view of the Rednecks' Mars Mission from the large screen at mission control in Daddy's garage.

★ ★ ★

When I put together the idea for the National Geographic Channel show *Rocket City Rednecks* one of the first ideas I had was that I wanted to do a simulated mission to the Moon or Mars. The mission had to be cheap and simple but would also give us some insight into an actual space mission. We decided to "go to Mars" because the red planet just captures everyone's imagination so well. So, I did some calculations and realized that if we were going to pull this off in a weekend then our spacecraft would have to be really really really really fast. If we could accelerate our spaceship at 2 g then we could walk around inside the spaceship as opposed to floating around in it. We'd be twice as heavy but we could manage it. And, it would make the mission much quicker. If we left at Mars' closest approach to Earth we'd have to travel about 10,000,000 km (10 million km). At 2 g constant acceleration we could cover that distance in about a day, but then we'd fly right past Mars at a whopping 22.8 million km/s! I decided to reduce the acceleration a bit so that we could accelerate to the halfway point for one day and then turn the ship around and start decelerating to slow down as we approached Mars.

This mission profile let us simulate going to Mars as a two-day trip there. Of course this technology doesn't exist yet but it will some-day. The point was to do a Mars mission simulation and learn some things. We've already learned how to go and not worry about the long trip in microgravity.

Once we were in the ship our plan was that we could never step out of the ship without spacesuits on. So, I picked my astronaut crew. I used my buddy Rog as our general "fix it" engineer and pilot. I chose Dr. Pete Erbach (my brother-in-law) as my science officer. And of course, this was my time to play Captain Kirk!

We loaded up our spaceship with food, water, and beer and were ready to go. We had found an old worn out RV that I purchased and we fixed it up to be our spaceship. We outfitted it with sensors, food storage, a bathroom, a water reclamation system to purify urine back into water, and we installed communications systems so we could phone home. As we approached Mars we also implemented a speed of light lag in our communications to make this a more realistic simulation.

For three days we were either in the spacecraft, doing EVAs on the craft to fix things or buy gas (which we used as a simulation of fixing something outside on the ship), or EVA on Mars, (which was an Alabama red clay dirt pit that looked a lot like Mars, actually). We built a shelter on the surface. We got tired of wearing those dang space suits. We had problems with the communications systems. We missed our wives and kids. And we drank pee.

All in all, it wasn't a bad simulation of a futuristic mission to Mars considering that we spent about $2,500 doing it. We learned how we could do the simulation better. We learned how we could implement simple concepts for using resources when we get there for things like shelters. We also got an inkling into the psychology of living in close quarters with our flatulent "fix it" man. It was fun and we learned a lot more than I was expecting to from the outset. I hope to do it again sometime on a bigger scale.

Toward that end, a multi-national experiment was recently concluded, known as Mars500. Some folks have actually done this red planet simulation on a much, much bigger scale.[16]

Mars500 was a joint undertaking by ESA's European Programme for Life and Physical Sciences (ELIPS) and Roscosmos' Institute of Biomedical Problems (IBMP) to accurately simulate the mental, psychological, and to a limited extent, physical effects of a manned trip to Mars. It encompassed a 520-day (seventeen months; nearly a year and a half) "mission" in a hermetically sealed habitat.

While this simulation could not be conducted in a microgravity environment, this is much longer than the typical six-month expedition aboard ISS, and thus provides data that has yet to be obtained. Meanwhile, such microgravity data is being diligently collected aboard the International Space Station.

The habitat itself, located in Moscow at the IBMP facility, consisted of four hermetically sealed interconnected modules comprising the ship, and one additional module representing the Martian surface, totaling 19,408 cubic feet (550 m³). After 250 days, the crew of six was

---

[16] http://www.esa.int/SPECIALS/Mars500/
http://www.esa.int/SPECIALS/Mars500/SEM7W9XX3RF_0.html
http://www.esa.int/SPECIALS/Mars500/SEMGX9U889G_0.html
http://esamultimedia.esa.int/docs/Mars500/Mars500_infokit_feb2011_web.pdf

divided in half; half stayed "on orbit," while the other half "landed on Mars." While on "Mars," they suited up and conducted several extended EVAs on the "Martian surface," using the Martian surface simulator. The region of Mars thus simulated was Gusev Crater, near the Martian equator and not far distant from where NASA's little rover Spirit conducted operations. This is an especially interesting area in that it appears to show evidence of water flow in the past, including sediments, stratification, and delta formation.

During the entirety of the mission they lived and worked like astronauts and cosmonauts on the International Space Station, performing experiments, doing maintenance on their "spacecraft," eating limited consumables, sleeping, and exercising. They worked a nominal seven-day week with two-day weekends, barring "emergencies," which were simulated. For the entire mission duration, they were monitored and their vital signs and psychological conditions recorded. Communications were realistic, and provided not only "interference" but time delays appropriate to the presumed distance from Earth. This latter ranged smoothly from 8 to 736 seconds, with the short times close to Earth, increasing to a maximum on flight day 351, when they were presumed to be at Mars, then reversing for the return trip home. Much of the time, therefore, communications occurred via e-mail or text, as this not only requires less bandwidth than video, but does not provide the confusion of a long video delay.

"When contemplating missions beyond low Earth orbit, such as to the Moon and Mars, daily crew life and operational capabilities may be affected by the hazardous space environment, the need for full autonomy and resourcefulness, the isolation, the interaction with fellow crewmembers and other aspects.

"A better understanding of these aspects is essential for development of the elements necessary for an exploration mission. Whereas research onboard the ISS is essential for answering questions concerning the possible impact of weightlessness, radiation and other space-specific factors, other aspects such as the effect of long-term isolation and confinement can be more appropriately addressed via ground-based simulations."[17]

---

[17] http://www.esa.int/SPECIALS/Mars500/SEM7W9XX3RF_0.html

★ ★ ★

Toward this end, candidates were selected. Like any astronauts, there were minimum selection criteria, which consisted of the following. Age: between twenty-five and fifty years old. Education: they each must have higher education in the form of post-graduate degrees. Professional requirements: the candidate had to be one of the following—a general practitioner having skills in medical first aid, a physician investigator or researcher having skills in clinical laboratory diagnostics, a biologist, an engineer specializing in life support systems, an engineer specializing in computer science, an engineer specializing in electronics, a mechanical engineer. Language skills: knowledge of the Russian and English languages at a professional level was required. There were also requirements as regards mental and physical health. Like any astronaut/cosmonaut/taikonaut candidate, the canadates underwent extensive medical testing. This winnowed a field of three hundred European candidates down to twenty-eight. These twenty-eight were invited to interview with the selection board. The twenty-eight became ten, who were subjected to extremely detailed medical and psychological testing equivalent to jet pilot qualification, in addition to in-depth interviewing to determine motivation and compatibility. Finally, four Europeans were selected and put through four months of rigorous training—including survival training. At the end of this time, the primary and backup teams were determined.

In its entirety, more than 6,000 people from forty countries applied for the 520-day crew.

The end result was that six crewmembers were selected—three Russian cosmonauts, two European astronauts, and one Chinese taikonaut. They entered an isolation chamber in June 2010, and emerged in November 2011. The six crewmembers, their home towns, and professions are:

- ✿ Romain Charles; Saint Malo, France; engineer
- ✿ Sukhrob Rustamovich Kamolov;
  Moscow, Russia; surgeon
- ✿ Alexey Sergevich Sitev; Star City, Russia;
  military naval engineer/shipbuilder
- ✿ Alexandr Egorovich Smoleevskiy; Moscow, Russia;
  military physician/physiologist

❊ Diego Urbina; Italy (home town not given);
   engineer/astronaut trainee
❊ Yue Wang; Beijing, PRC;
   teaching/environment adaptability studies/taikonaut trainer

Their ages range from twenty-nine to forty. But they were not simply thrown into the long-duration simulation and allowed to fend for themselves.

In November 2007, a fourteen-day study was carried out. This adequately tested the habitat and operational procedures. And from March to July 2009, a 105-day study was conducted as a preliminary to the full "mission." With both successful, the European crew for the "Mars mission" was announced on March 23, 2010, with the Russian complement being announced on May 18. On June 3, 2010, the complete 520-day simulation commenced.

Simulated milestones were as follows[18]:

❊ 3 Jun 2010: Beginning of experiment—hatch closed, lift off
❊ 15 Jun 2010: Undocking from orbital assembly laboratory
❊ 23 Jun 2010: Transfer to heliocentric orbit toward Mars
❊ 24 Dec 2010: Shift to spiral orbit toward Mars
❊ 1 Feb 2011: Enter circular orbit around Mars
❊ 1 Feb 2011: Mars Lander hatch opens
❊ 8 Feb 2011: Completion of loading, Lander hatch closure
❊ 12 Feb 2011: Undocking, landing on Mars
❊ 14, 18 and 22 Feb 2011: Egresses (EVAs) on Martian surface
❊ 23 Feb 2011: Ascent, beginning of quarantine
❊ 24 Feb 2011: Docking with interplanetary craft
   (command module)
❊ 26 Feb 2011: End of quarantine
❊ 27 Feb 2011: Habitation module hatch opens
❊ 27 Feb 2011: Crew transfer to Habitation module;
   begin transfer of trash to Lander module
❊ 28 Feb 2011: Lander loading with trash ends
❊ 1 Mar 2011: Hatch closure, Lander undocking
❊ 2 Mar 2011: Enter spiral orbit away from Mars
❊ 7 Apr 2011: Transfer to heliocentric orbit towards Earth

---

[18] http://www.esa.int/SPECIALS/Mars500/SEMGX9U889G_0.html

❦ 15 Sep 2011: End of communications delay,
   switchover to voice communications
❦ 13 Oct 2011: Shift into spiral orbit towards Earth
❦ 4 Nov 2011: End of 520-day study, crew lands on Earth

By way of comparison, the longest Apollo mission was Apollo 17 at approximately twelve days from launch to splashdown. The longest Space Shuttle mission was STS-80, outfitted with the Extended Duration Orbiter (EDO) pallet, at a mission elapsed time of seventeen days, fifteen hours, fifty-three minutes, and eighteen seconds. The longest Skylab habitation was the SL-4 mission at eighty-four days, and the typical ISS expedition length is six months (~180 days). The record duration in space is currently held by cosmonaut Valeri Polyakov, who spent 437 days, 18 hours (>14 months) aboard Mir, in 1994–95.

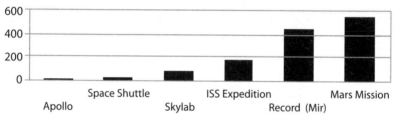

But the average estimated round trip flight time to Mars with current technology is 400 to 450 days, and that does not count the time spent by a lander on Mars. The time is partly determined by the locations of Earth and Mars relative to each other in their orbits, and partly by the amount of thrust available to eject the spacecraft out of Earth's heliocentric (Sun centered) orbit and into Mars' higher heliocentric orbit. There is a "resonance cycle" every twenty-six months that enables a relatively low energy trip to take place. Missions would be planned to take advantage of these resonances. This is *NOT* the shortest distance between the planets, nor the time of closest approach. It is known as a Hohmann transfer, and is an elongated ellipse, tangent to each planet's orbit at opposite sides of the ellipse, and would in fact require launch from Earth when that planet is on the opposite side of the Sun from Mars. The estimated flight time of 400–450 days is based on this orbital transfer planning method.

If the launch occurs at a different location in the resonance cycle, or the lander delays too long on Mars and thus misses the resonance window, the flight duration could take much longer as a result of the increase in energy requirements and the necessity to eventually synchronize the spacecraft with the planet(s) involved. Likewise, if a different approach than conventional rockets' "point, shoot, and drift" means were used (e.g. solar sail, ion propulsion, nuclear engines) then the duration could very well be reduced by taking advantage of a constant acceleration instead of allowing an initial push, gravity, and orbital mechanics to do all the work. But none of these technologies are mature enough to use immediately. It would take a decade of testing to have a viable alternative. The most likely candidate would be to use a hybrid of a nuclear thermal rocket for high thrust and then a nuclear electric propulsion system for long duration but lower thrust. We could do this with modern technology if we just invested the money to build and test it.

Given the difference between the typical orbital stay and the duration of a Martian mission, the Mars500 experiment assumes even more importance.

The facility in which the extended experiment was conducted was specially built at IBMP. Inside the building was not only the series of interconnected isolation modules but also the control center and offices. As before mentioned, there were four hermetically sealed compartments and a module to simulate the Martian surface. These four compartments were the medical module, the habitat, the lander, and the stowage.

The medical module was 10.5 x 39 feet (3.2 x 11.9 m). It contained a toilet, two medical berths, routine medical equipment, laboratory and diagnostic equipment, and telemedical capabilities. With physicians aboard, should anyone have become injured or ill, they could have been treated in the medical module—even quarantined—without the necessity of breaking the isolation.

The habitat was the main living quarters. It was 11.8 x 65.6 feet (3.6 x 20 m) and held six crew compartments, a combined kitchen and dining area, a toilet, the living room, and the control room (flight deck). Each crew cabin had a floor area of between 30 and 34.4 square feet (2.8 to 3.2 m^2), and contained a bed, desk and chair, and shelving for the astronaut's personal kit.

The lander module was only used during the actual Mars landing phase of the mission, which lasted thirty days. It was 20.7 x 20.2 feet (6.3 x 6.17 m) and held three open bunk berths for crewmembers, a toilet, two workstations, telecommunications equipment, a flight control and data collection station, as well as air conditioning and ventilation; gas analysis, sewage, and potable water systems; and a fire alarm and suppression system. It also had two access tunnels, one for ingress from the orbital "spacecraft," the other for egress to the "Martian surface." This latter included an airlock, in which were stowed the Russian Orlon space suits used for the EVAs.

The stowage module was 12.8 x 78.7 feet (3.9 x 24 m) and divided into four compartments, depending upon use:

Compartment 1—refrigerated perishables (foodstuffs)
Compartment 2—nonperishable foodstuffs
Compartment 3—experimental greenhouse
Compartment 4—bathroom, sauna, gymnasium

The Martian surface module was . . . large, with a volume of just under 42,400 cubic feet (1,200 m³). It was accessed via the airlock from the lander module, and crewmembers entering it were required to wear space suits at all times while in its confines.

What makes the Mars500 experiment different from an extended stay on Mir or ISS? Why is it so important to simulate a Martian mission? What are the dangers aside from those due to the interplanetary space environment? What makes this any different from the ISS, or a stay at an Antarctic research station?

For one, the ISS has an escape craft. Even should that craft not be attached or be disabled, help is not far away. The same goes for one of the Antarctic research stations. We only have to look at the physician with cancer some years back, who was emergency airlifted from the base when the medical facilities there proved inadequate to her treatment. Yes, it was dangerous, but it was possible. A Mars mission may be many millions of miles away. Mars at its closest approach to Earth is still over 6.2 million miles distant, and as mentioned, this is not the optimal time to launch a spacecraft. The next minimum energy trajectory window would not open for another two years and two months after the initial mission was launched, and then would have to catch up to the first. Even should it do so, there is still the return trip to Earth. An emergency situation would be long past before a rescue could be effectively mounted.

And there are many potential emergency situations that involve merely the crew. These include physiological, psychological, and emotional dangers. Experiments on many of these were conducted during the Mars500 simulation.[19]

"Adaptation, group structure, and communications of confined and isolated crews" was one such scientific "protocol" that was conducted. This looked at some factors involved in both individual and social adaptations to the isolated, confined environment. How many times do we, when we need a break, go for a walk? What would we do if we were unable to do so? How would we handle the stresses? How do we deal with seeing the same people, the same four walls, the same activities each day, with no change and no outside influences? How does the human process, analyze, and adapt to this situation? This was the nature of this study.

"Association between psychological and cardiac functioning in a confined population" was another study. It is a known fact that forced inactivity (as well as extended duration microgravity) causes cardiovascular deconditioning. The concern is that the confinement could also lead to emotional stress which could then in turn stress the cardiovascular system. This study was aimed at monitoring both emotional stress and cardiovascular behavior and attempting to correlate the two.

"Clinical-physiological investigations for maintaining physical fitness and body composition by resistive vibration exercise" was the third experimental protocol. Using special equipment, it was an attempt to maintain the posture and locomotor systems during confinement. Comparisons were made before, during, and after the confinement to determine efficacy.

"Development and testing of an operational tool for learning, training and maintenance of space specific complex skills for object hand control with six degrees of freedom" was the name of the fourth investigation. Software led the participants through increasingly difficult coordination tasks before assigning them a highly complex spacecraft docking simulation. This helped determine the degree of

---

[19] http://esamultimedia.esa.int/docs/Mars500/Mars500_infokit_feb2011_web.pdf

hand-eye coordination loss and attempt to overcome it, so that manual docking with the landing module at Mars would not be rendered impossible.

"Evaluation of Stress and Immunity" was another experiment. Previous long-duration studies have indicated that there is a decrease in the effectiveness of the immune system during microgravity or long confinement. It has further demonstrated that stress is a contributing factor. This investigation attempted to quantify that effect and look toward a possible pharmacological remediation.

"Effect of blue-enhanced light on alertness and sleep-wake behavior" studied the effects of light from various parts of the spectrum on the crewmembers. This is well known to those who suffer from Seasonal Affective Disorder, or SAD. The lack of certain frequencies of light in artificial lighting, especially in an individual who has no opportunity for direct sunlight for extended periods of time, leads to problems on Earth, let alone in a crew module in space. It can induce sleep irregularities, depression, mood alteration, alertness issues and more. This experiment, therefore, enriched the blue frequencies in the cabin lighting, and looked at its cumulative effects.

"Effects of group dynamics and loneliness on cognitive and emotional adaptation to extreme, confined environments" was next on the list. The primary question here was, "Does long-duration space flight create stronger stresses than normal on the human psyche (e.g. loneliness), and if so, how does this affect performance and professional conduct?"

"Influence of physical activity and dietary supplements on the serotonergic system and its implications on performance and mood." Serotonin is a neurotransmitter (it passes between neurons in the central nervous system) which is derived from tryptophan, and regulates some important bodily functions, including mood, appetite, sleep, memory, and learning. It is why such foods as milk and turkey, both high in tryptophan, tend to make us sleepy—as it boosts serotonin levels, we become content and drowsy. Its lack can lead to chronic, and even severe, depression. (Many modern antidepressants are based on the concept of preserving serotonin in the nervous system.) This was an in-depth study in order to determine if food and supplementation could help have a positive influence on behavioral patterns, including mood, sleep, and motor skills. The experiment made use of supplementation with either tryptophan or branched chain amino acids, both of which are important to a healthy diet.

A very important study included the "medical skill maintenance during long duration spaceflight." This was an effort to determine if both hands-on and mental medical knowledge and skill was lost during an extended mission. Since astronauts by and large are selected partly on their health, while a physician aboard is essential, he or she may not have anything to do for long periods of time. The old adage, "Use it or lose it," comes into play here. The study not only measured the physicians' capabilities at various points in the flight, it also utilized various kinds of refresher courses administered periodically, and determined their efficacy.

"Microbial ecology of confined habitats and human health" is definitely important. We know that microbes and virii mutate at a rate much faster than other, more sophisticated organisms. What would a microbial colony, isolated from other microbes, do over a long period of time? Do microbes become more or less virulent? Multiply faster, slower, find new niches to inhabit? Do the natural microbes present in the human digestive tract continue normal behavior? To this end, certain areas were treated with antimicrobial compounds to prevent contamination, and those areas monitored. In addition, the crew were fed probiotics to determine their effects upon the internal flora and fauna.

The Mission Execution Crew Assistant is not so much an experiment as an attempt to develop a machine-human interface for complex and emergency activities. According to ESA, it "will empower the cognitive capacities of human-machine teams during planetary exploration missions in order to cope autonomously with unexpected, complex and potentially hazardous situations. The goal is to encourage human and machine groups to act in a distributed, autonomous but cooperative way."[20] If successful, this could lead to an actual interactive ship's computer not unlike those normally found in science fiction.

The "Neuro-Immuno-Endocrine and metabolic effects of long term confinement" experiment intended to verify the effect of long-term

---

[20] http://esamultimedia.esa.int/docs/Mars500/Mars500_infokit_feb2011_web.pdf

confinement and isolation on metabolic and gonadal function, with the ultimate goal to develop hormonal and nutritional countermeasures to the negative impacts.

"Omega-3 polyunsaturated fatty acid and psychological wellness during long-duration space missions" is an interesting study, as a deficiency in polyunsaturated fatty acids has been found to be a contributor to inflammation and depression. The study will attempt to determine what levels are necessary in the blood in order to maintain a healthy, normal function, especially under trying circumstances such as a long space flight would produce.

The last study was somewhat esoteric, but no less important for all that. "Personal values on missions to Mars—implications for interpersonal compatibility and individual adaptation" will help determine how personal belief and value systems influence the individual and sustain him or her on a long space flight, making the crew member more or less resilient, as well as how such things—or rather, their influence on interpersonal behavior—affect the cohesiveness and compatibility of the crew.

The data for all of these has now been collected. The only thing that remains is analysis and conclusions. In addition, Roscosmos is considering conducting a similar experiment aboard ISS in the near future, where full microgravity conditions will exist. This may occur after 2014, and involve at least two astronauts for a minimum of 18 months.[21]

One of the studies not performed is the effect of mixed gender crews. Although the earlier, shorter duration simulations had mixed crews, the 520-day simulation had an all male crew. So this aspect was not addressed in terms of potential relationships. The human psyche being what it is, it would be expected that some pairing off might occur during a lengthy trip, if it has not already happened during training. How, then, do the stresses of the trip, the isolation, affect the relationships? Do they become stronger, do they aid those involved in the relationships to adapt better by alleviating loneliness and similar stresses? Do they become weaker because the continual contact leads to burnout?

---

[21] http://www.space.com/13500-mock-mars-mission-mars-500-ends.html

And what about any crew members who may not have a partner? How would jealousy affect interpersonal crew relations? Would jealousy even occur? These are matters that should be investigated.

My biggest question is probably, "Why wasn't the United States involved in this simulation?" Every other major country capable of manned space flight was. Heck, why didn't the United States *do* this simulation? There was some great science that came out of this study, and it wouldn't have been that hard to do, or even been that expensive. The total cost for the project was initially estimated at only US $20.5 million[22] at the outset, and according to SPACE.com, ended up being only US$15 million.[23]   Unbelievably, they actually came in under budget. Of course, their budget was nowhere close to the $2,500 and weekend-long experiment that we did on *Rocket City Rednecks*, but I'm sure they had just as much fun as we did. I hope they packed enough beer in that thing.

[22] http://suzymchale.com/ruspace/mars500.html

[23] http://www.space.com/13500-mock-mars-mission-mars-500-ends.html

CHAPTER 10:
LET'S GO TO THE RED PLANET

The red planet Mars has always captured our imagination. It is nearby. It is not too different from Earth. And science fiction writers have been enamored by it for a century or so. In fact, one of my own SF novels is based around action on Mars. Hey, everybody else was doing it.

As a kid I used to look at it through binoculars and my telescope and wonder about that ice cap. Is that really dry ice or is some of that snowy looking pole water ice? We need to go there and find out for sure! If we did plan to go there it would be extremely difficult and dangerous. Mars is nearby in a solar system sense, but as far as manned rockets go it is very very far away.

We would likely have to land there and plan to stay for a while in order to gather water or other materials for fuel for the return trip. So that means living on Mars for some extended amount of time. Simply staying in the landing vehicle like on the Apollo Moon missions would be uncomfortable. It is okay to camp out for a weekend or even a week in a camper sized compartment but not for months and months.

This is where the *Rocket City Rednecks* comes in. We talked about doing a Mars simulation mission but we also talked about how would you build habitats and other infrastructure from stuff that is already on Mars. Our thoughts were to use craters and caves that might already be there. Then use some lightweight but strong material to cover it.

Pete is a big fan of spray foam insulation/adhesive and I'm becoming one myself. He convinced me we could build a dome using just spray foam. So, we had Michael dig us a crater in the dirt that we could use as a mould and then we filled it with foam. A lot of foam!

Once the foam dried we flipped it over and covered the crater with it. Now if we'd had time on the show we would have shown that if you were really on Mars, as soon as you pressurized the compartment to an Earth atmosphere you'd blow the foam lid off the crater. The next step to prevent that would be to use Mars dirt and cover it or to throw ropes over the dome and tie it off to spikes in the ground. Perhaps a bit of both ideas would be good.

We had planned to go into more detail with the dome and to add lights and cots and make it more livable, but we had plenty of footage for a thirty minute show already. Maybe one day we will get to revisit the idea and go into a little more detail. Heck, hopefully, one day before long we'll get to see some astronauts go there and do it! Now that would be great to see some real rednecks on the real red planet!

List for a Self-Sustaining Mars Colony, the Rocket City Redneck Way

One of the big Mars mission proposals, the 100-Year Starship concept, is not very well thought out from the get-go. That is a severe case of putting all your eggs in a very leaky basket.

The right way would be to start right now sending the infrastructure to Mars.

Medical	Quantity, Etc.
CNC machines	37
X-ray/MRI machines	37
Each drug known to man	One five-gallon bucket each
Every medical instrument known to man	Two each
Band-Aids	Enough to choke an arena full of mules

Power Sources and Construction Tools	Quantity, Etc.
Solar cell farms	Big ones, lightweight for space travel
Fission reactors	Four or five NASA SAFE 300-sized reactors would do.
Bobcats and Caterpillars retooled to run on electric power and in the harsh Mars environment	½ rocket full. We should be building and testing these now
Hammers, wrenches, shovels, drills, nuts & bolts, screws, augers, saws of every style, welders, knives, files, etc.	The other ½ rocket full. Must have a machine shop. A Navy aircraft carrier shop might be a good place to start.
Spray foam	A bunch, need to invent some that we know will work on Mars
Chicken wire	1,000 yards
Duct tape	10,000 yards
Every adhesive known to man	100 gallons each
More duct tape	Nooks and crannies in the rocket stowage

Living Items	Quantity, Etc.
Microwave ovens	Five
Camping chairs	42

Living Items	Quantity, Etc.
Rechargeable batteries of all shapes and sizes and voltages OR everything we send is designed for one battery type (makes more sense)	One battery type, half a rocket full.
Big-screen TVs and a special pipeline to get the SuperBowl, Iron Bowl, BCS Bowls, and all SEC football games streamed	A 1,000 terabyte drive full of movies and TV wouldn't hurt
More duct tape	As much as will fit
HDPE cylindrical storage storage tanks	As many as we can manage
Structural aluminum tubing and angle pieces	Two full missions with nothing but
Every system component intended to be used	Two full missions with nothing but
Clothing	All sorts
Clorox wipes and toilet paper	Rocket full
More toilet paper	Cram it in!
Masonry tools, concrete mixers	Four
Automatic sandbaggers	Four
More spray foam	A gazillion gallons

Living Items	Quantity, Etc.
Pure grain alcohol	Two tons, and maybe a bunch of red SOLO cups
Aerostats w/helium	Deployable cell towers, recon, etc.
Rope (lots of it)	Tested for Mars environment
Zip ties, all sizes	Two tons
Paper clips	Two tons
Extension cords	Rocket full
Plastic ziplock bags	Two tons
Vacuum bagger and bags	Tons of bags

Surface Movement	Quantity, Etc.
Battery-adapted Polaris electric ATVs	Ten
Personal space suits	Four each
Oxygen farming/collecting equipment	Enough for all occupants of colony with contingency

Food and Water	Quantity, Etc.
Every seed/seedling known to man	Five tons
Compost, peat moss, and manure	Five tons each

Food and Water	Quantity, Etc.
Composition 4. We can make dry ice bombs once we find water. There is plenty of dry ice on Mars.	One ton
PVC pipe, couplers, fittings, and glue of all sizes and shapes	Ten tons
Extremophile enhanced kudzu	One sprig will be enough. And other possible plants that could live on Mars
Cold weather algae and arctic microbes	Ditto
Cockroaches (preferably the largest edible variety)	100 lb, eggs. Nasty sounding so astronaut/colonists must live by one rule: *Never think about what you are eating. Just eat.*
Maybe a bunch of flowers and honeybees	100 lb, eggs. Have to pollinate the plants
Worms and grubs! You can eat those too	100 lb, eggs.
Potatoes and other veggies like them would not go bad just lying out on the Martian soil. They might actually grow at the equator in summer. It does get sixty-seven degrees at times only a few inches from the surface.	One ton each variety; absolutely *no* legumes or cruciferous veggies!

Food and Water	Quantity, Etc.
Frozen/vacuum sealed meats and an electric barbecue grill	Two tons
50,000 jars of peanut butter, again, just stored out on the surface. This would actually be a good Bell jar experiment.	The trapped air might explode the jars so we may need to invent new harder containers for it.
Granola/Powerbar/ Snickers bars etc.	50,000
Metamucil and lots of it until the plants start growing	Lots
More toilet paper	Start cramming it in stowage space.
Powdered Gatorade	Two tons
Compressed sugar	Two tons
Other long term keepables like microwave popcorn, rock candy, spices, flour, etc.	Two tons
MREs	Five tons
Water to last until we find a good source. There should be plenty at the edge of the ice cap.	A few tens of tons

Science, Engineering, and Communication	Quantity, Etc.
Communications equipment (planet-to-planet and local RF radios and repeaters)	Lots!
Metal foundry (there is bound to be plenty of ore on Mars)	Two
Science equipment of all sorts	Lots and lots!
Enough fireworks for 100 4th of July celebrations	Shipped in separate rockets
Personal effects	One rocket full
Foxfire books, military survival manuals, field medical texts, etc.	Three copies each

We send all this and. . .

Kitchen sinks	14

. . . to Mars over the next ten to twenty years. If we want to send some robots to start stacking stuff and digging holes, that would be okay, too. Oh, and we should send a big basket of eggs just for fun.

Then, and only then, do we build a slimmed-down, fast-as-we-can-build crewship that we fire off. I'm thinking a hula hoop that spins to keep the crew from wasting away (other than throwing up all the time) with a central Mars Surface Access Module (MSAM) hub vehicle.

We eject the hula hoop when we get there and let it crash into Mars (more raw materials). It should be refuelable, and have an ascent engine. Next we'll send a return ship to orbit the planet but that might take a few years.

We could do a real study of this and decide an exact manifest of items and how many rockets it would take and how much that would cost. Say we send five Delta 4/Atlas 5 heavy rocket packages per year for ten years—that is fifty rockets. Each of these launches will cost about $200 million or $1 billion per year for five missions. Over ten years there is $10 billion spread out on just the launch costs. And, each of these launches could only carry about 4,000-5,000 lb to Mars. That is 25,000 lb per year of stuff. This would give us 250,000 lb of stuff sent to Mars in ten years. Maybe that is enough stuff. Maybe.

The development of each of the components is the expensive part. All of it must be flight and space ready. It would take at least another $3–5 billion per year to build, test, pack, and load all this stuff. So that is another $30–50 billion to the total ten year budget (not that bad actually). To send all our stuff would cost us about $60 billion dollars over ten years.

Then, there is the manned spacecraft! $10 billion for that is probably not a wild guess, but could be off by a factor of two or three simply because we've never built a manned Mars ship before. It is really difficult to say at this point. But, the ship would be being developed in parallel with the supply missions.

We could do this. We could put a real self-sustaining colony on Mars in ten to fifteen years if we started now. Total budget is somewhere between $70–$100 billion.

We could also continue to send supplies as needed.

★ ★ ★

All right. So we're planning a trip to Mars. We build a honkin' big rocket, put a capsule and lander on top, point it at Mars and fire—right?

Not quite.

Let's talk about some basic things first. Like gravity around two objects orbiting a common center of mass, and things like that. We'll start by talking about the Moon and the Earth.

To explain orbits and gravity and such, scientists often use the analogy of space (actually space and time together but that is over complicating the discussion) being like a stretched rubber sheet. Using this analogy (it was actually Einstein's), any massive objects create dents in the sheet. Imagine setting a bowling ball on a trampoline. The heavy ball will create an indention in the trampoline's surface material. If you rolled a small tennis ball across the trampoline slow enough, it would fall in the indention until it landed on the bowling ball. If the tennis ball is rolled fast enough it will keep on rolling by the ball but the tennis ball's course will be altered. Some really interesting things happen when those dents are moving, especially if they're orbiting each other, something interesting happens to the rubber sheet. It forms shapes that remain constant relative to the orbiting objects.

Let's back up a minute. Note I didn't say one orbits the other. That's not what happens. They orbit a common center of mass, which is technically called a barycenter. Simply put, it is the point in space where the entire mass of a system may be put and all outside actions and reactions will remain the same, and it is the point around which the system revolves. In a lot of instances, that center of mass might be located *INSIDE* one of the objects, but it won't be at the exact center. In some instances, it won't be inside an object at all.

Now, back to our stretched rubber sheet. Let's say we have a couple of marbles on it. One is a shooter—the Earth. The other is a duck, a regular marble—the Moon. And they're rolling around and around their common barycenter. We discover by putting a tiny BB (pretend that's a spacecraft or a satellite) on the rubber, that there are some unique bends that form in the rubber, and stay put relative to the marbles. If we put a BB in one of five places, it will tend to stay put there, regardless of the movement of the marbles, and it will follow the marbles around as they revolve.

These points are known as Lagrangian points, or libration points. There will be one between the two marbles, and one on the back side of each marble (assuming the front side faces the other marble), and one out to right and left of the marbles. They are points of gravitational stability in the orbiting system. You really can put an object there and, within certain parameters, they'll stay put. If we only had to contend with the Earth and Moon, they'd stay put indefinitely. But we also have the Sun and other planets to contend with, so in reality we'd have to tweak our position occasionally. Now let's look at each point separately, because they're not quite shaped the same.

If we draw a line between the centers of the two marbles and extend it outward, we'd find that three of those points lie along that line—one in between, and one on either side of our marble system. The one in between is called the L1 point. The L1 point is always located where the gravitational pulls of the two bodies are exactly equal. If you nudge the BB out of the line between the marbles, it will roll back into the L1 point—it's like the "hammock" between the dents in the rubber. But if you nudge the BB a little way toward either of the marbles, it will start rolling downhill toward whichever of the marbles is closest.

As it happens, every time an Apollo spacecraft went to the Moon, it passed through the L1 point. The figure-eight trajectory of the total out-and-back flight has its crossover at the L1 point. Would it be possible to get to the Moon without going through the L1 point? Maybe, but why would you want to? It takes just as much energy to climb out of Earth's gravity well (the dent in the rubber sheet) there as anywhere else, but once you do, and once you pass the L1 point, you have the advantage that the Moon's gravity immediately pulls you in, and all you have to do is correct your course into an orbit instead of a crash landing. If you came out of the gravity well anywhere else, you'd have to change direction and head for the Moon, and burn enough energy to get caught in the Moon's gravitational well, and then correct course. That's extra fuel burned. Extra fuel burned means extra weight carried up from Earth's surface. Extra weight makes it harder to get out there. So, we use the L1 point as a means of saving energy in a trans-lunar injection.

L2 will always be on that same line, but on the outside of the smaller body—in our example, the Moon. And L3 is on the same line, but outside the larger body—Earth. They're similar in nature, so we'll talk about them together. The L2 and L3 points are the points where the combined gravity of our two objects just equals the centrifugal force on our BB, or in reality, our spacecraft. And just like the L1 point, if you try to nudge it away from that line between the object centers, it'll tend to roll back. But if you nudge it toward the nearest object, or away from it, it will tend to roll away from the L point—either toward the nearest object, or out into space.

Now, the L4 and L5 points are identical to each other, but different from L1, 2, and 3. They are pretty much interchangeable, and are sometimes called Trojan points, because the first demonstrable proof of their existence came with the discovery of the Trojan family of asteroids in the asteroid belt. There are two areas in the asteroid belt where there are permanent clusters of asteroids that just stay there. It so happens that those two areas correspond to the L4 and L5 points of the Jupiter/Sun system. (And yes, it turns out that Earth has its own Earth/Sun Trojan asteroids—meaning they're reasonably close by, and might be a good target for an asteroid mission. More about that later.) The next step is to draw a triangle on our stretched sheet of rubber with some chalk. We draw a line between the centers of mass of our shooter and our duck (the Earth and the Moon). Using that as the base, we draw an equilateral triangle with a vertex pointing in the direction of revolution, and another one pointing opposite it. These vertices are the L4 and L5 points, and they will be sixty degrees ahead and sixty degrees behind our orbiting system.

The reason these two points are stable is that the distance to the two masses are equal. This causes the resulting sum of gravitational forces to act as if it were the barycenter pulling on the spacecraft. At this point there is an equilibrium produced because, essentially, anything at L4 or L5 is orbiting the same barycenter as the Earth and Moon. One thing to keep in mind is that we're talking about tiny objects like spacecraft relative to the planet-sized objects. However, L4 and L5 points are like a bit of a hill in our stretched rubber. Nudge the spacecraft in any direction and it will tend to roll away—either in towards the system, or out into space.

Putting any object in any Langrange point is like parking a trailer in a lot, provided you've got a little onboard thruster and orientation system to keep it nudged into the center of the point—you can go off and leave it for a day, or a week, or a year, and come back and it will still be there. These Lagrange points are really good places to park things like space stations, fuel depots, supply depots, asteroid busters, cargo coming inbound to Earth, cargo shuttling to and from the Moon, all kinds of useful things like that! Science fiction writers have been using them in stories for decades for just such purposes.

Now, so far we've simplified the thing and only looked at it in two dimensions—the dimensions of a rubber sheet. Let's expand our thinking; as Mr. Spock once said, "He is intelligent, but not experienced. His pattern indicates two-dimensional thinking." We mustn't fall into the same trap that Kirk used to end Khan.

So, we will also find that there are orbits that can be created in and around the Lagrangian points, and about the line through the centers of masses.

Halo orbits are just what they sound like, ellipsoidal orbits that occur out of the plane of the two-body system, but rather orbit the center line near one of the L points. A sufficiently large halo-orbiting comm satellite, placed near the L2 point, for example, could enable the entire lunar farside and the Moon-facing Earth side to send and receive signals virtually continuously.

Lissajous orbits are quasi-periodic, three-dimensional orbits that meander around L points. Think of them as 3-D oscilloscope patterns! These types of orbits are wave functions that travel in a circle or ellipse and eventually return to the starting point, then repeat. They are complicated looking and may even appear to have little pattern to them, but they are a repeating wandering pattern.

Lyapunov orbits are similar to Lissajous orbits except they lie in the plane of motion of the two-body system. These are other potential "parking lots" in space.

Any of these orbits can be utilized in our future spacefaring, for many things, including shuttling material to and from the Moon and Earth, communications, supply locations, drop-off points, and orbital construction projects. In fact, the best thing to do for a Mars mission (or farther) would be to construct the spacecraft in one of the L points, possibly even using materials mined—or even fabricated— in a lunar base, and shuttled up to L1, then out to another L point. L2, on the opposite side of the Moon, might even provide an excellent spacedock. Our spaceships could shuttle materiel up directly from the lunar farside, build it at L2, then time departure correctly to use the centrifugal force as an additional "push" along to Mars.

Now, let's go back and look at that figure-eight trans-lunar trajectory for a minute. It turns out that this figure eight is kind of important. We take that figure eight and rotate it around the line through the centers of mass of Earth and the Moon. The result is sort of an hourglass shape with the half of it on the Moon side being a little smaller than the bulb on the Earth side. Now, imagine that shape in the space between the Earth and the Moon, and imagine that everywhere on that hourglass's surface, the force of gravity is the same. Also realize that the gravity along this surface balances the centrifugal force (the force the spacecraft is exerting to pull away from the Earth), then what you have is called the "Roche lobes" of the system. Each one of the orbiting objects has a teardrop-shaped Roche lobe that is one "half" (they may not be the same size) of the hourglass. The hourglass moves with the two bodies, and spins about the barycenter like an hourglass used for a game of spin-the-bottle.

Now, the Apollo trans-lunar trajectory didn't really follow the Roche lobes. The Apollo flight did have one thing in common with the Roche lobes for the Earth/Moon system though, the intersection point is the L1 point in both cases.

What's important about the Roche lobe? Well, inside the Roche lobe, the spacecraft belongs to the gravitational field of the central object. Outside the Roche lobe, not so much, and it is much easier to escape from the object's gravity. Low Earth orbit is well within Earth's Roche lobe, for instance. Anything that passes through the L1 point goes from one Roche lobe to the next, and anything that passes through a Roche lobe with a velocity that is less than the escape velocity of the object will be captured by the object. But anything passing through a Roche lobe with a velocity greater than escape velocity will still be affected. For instance, Comet Shoemaker-Levy 9 passed through the Roche lobe of Jupiter on a previous orbit, and this is what tore it apart into the "string of pearls" comet. The Roche lobe is close enough for the gravity of the central object to affect a large object by different amounts on different sides, so the side of the comet nearer Jupiter was pulled on harder than the side of the comet away from Jupiter, and the comet tore apart.

So, when we're doing all of our exploration out there, it's important to know where we are in the gravitational fields, and make sure we're going the right speed for where we need to be. But more so than just being careful, we need to be smart in developing our space plans to take advantage of such phenomena. We most definitely should start practicing at placing spacecraft at Lagrange points and keeping them there for long periods of time. The Lagrange points should be way stations with fuel, food, water, spare parts, and anything else we can think of that we might need along the way in space. Once in space, it is easier to go to a Lagrange point for stuff we might need than to land on a planet and then launch from there again. We need to be planning our first depot in space, and the Earth-Moon L1 point would be a great location.

We also need to put a station at the Sun-Earth L1 location. This station would be used to warn us of solar flares and coronal mass ejections that are extremely hazardous to our communications satellites, infrastructure, and power grids.

As we plan our missions to Mars and the outer planets we should consider any and all of the Lagrange points as useful storage depots, way stations, and even outposts. The Roche lobes are a little more difficult to determine a great application for other than to understand more about a complex two-body gravitational system. That said, though, if a spacecraft is placed anywhere on the Earth's side of the hourglass surface it will mostly stay on that surface without outside intervention or forces. But, if we push that spacecraft just right until it passes through the neck of the hourglass at the Earth-Moon L1 point it will then be on the Moon's side of the surface and will be under lunar influence. This is something we need to practice with in learning how to pass ships back and forth between the Earth-Moon sides of the Roche surface with a minimum amount of energy expenditure. There is likely to be a technique we have yet to implement that would reduce the amount of rocket fuel needed to transfer spacecraft from one part of the Roche lobe to the other.

It is concepts like the Lagrange and Roche locations that will require more spaceflight experiments than can be done on limited shots to the Moon or Mars. Only after we have created a long-term presence in space will we start truly being able to implement these types of concepts. We need to build flyers that wander around on these types of orbits for years. This would allow us to study how better to implement the attributes of the space around us. There might be new propulsion concepts like solar sailing, laser sailing, ion propulsion, or something else where implementing both Lagrange points and Roche lobes are to our great advantage. We will never know if we don't start spending some research money on advanced propulsion flight experiments.

Getting back to what Apollo did to get to the Moon was a much more simplified approach. The Apollo trans-lunar injection was a Hohmann transfer. And that's the same type of orbital transfer we'd probably use to go to any of the outer planets. Getting to the inner planets is a little more complicated, by going in the reverse order, but then they don't have environments that are very encouraging for manned missions anyway. Equatorial surface highs on Mercury are around 800°F (427°C, 700K) with only a trace atmosphere, and Venus isn't any better, with a mean surface temperature of 863°F (461°C, 735K) and a surface atmospheric pressure of 1,348 psi (93 bar, 9.3 million Pa). Landers on Venus typically don't last longer than an hour or two on the surface.

A Hohmann transfer is a transfer from a lower orbit to a higher orbit that uses the minimum amount of energy necessary to accomplish the maneuver. Remember about energy? More energy equals more fuel equals more payload equals more money and difficulty. It isn't necessarily always the fastest way, but it *is* one of the most direct of the energy-efficient transfers. The transfer in reverse is how to go from a higher orbit to a lower orbit, and in either direction, the energy requirements are the same.

It is usually only half of an elliptical orbit each way. And to go to a higher orbit, the spacecraft ignites the engine to provide additional thrust in the direction of orbital motion. This increases the speed and pushes the spacecraft out into a higher orbit. The final height of the orbit is determined by the speed increase, which in turn is determined by the engine burn. Then it simply coasts, following Kepler's Laws of orbital mechanics. At the apoapsis (farthest point) of the transfer orbit, the spaceship flips around backwards and ignites the engine just long enough to stabilize the ship in the new, higher orbit.

In the case of transferring from low Earth orbit to the Moon's orbit, or Earth's solar orbit to another planet, the whole thing has to be planned just right so that the Moon or planet arrives at the apoapsis of the Hohmann transfer just as the spaceship arrives at the same place. It's like leading the clay pigeon with a rifle, or a quarterback leading his receiver. This is where the concept of a "launch window" comes in. The launch window is that period of time during which the launch *must* occur for the alignment to be right for the transfer. In the case of Mars, they happen roughly every 1.67 Earth years. Then the reverse burn is used to match speeds and allow the other celestial body to capture the spaceship in orbit (inside that Roche lobe).

Getting back is a tad bit more complicated. The first thing you have to do on leaving Mars to come home is to use a Hohmann transfer to get up to the edge of Mars' gravity well! Then you kick out of the Martian gravity well, and do a retro burn to drop out of Mars' solar orbit (the ship is still moving forward in its own, more elliptical solar orbit), making sure you've timed everything right for the Earth to be there when you arrive at Earth's solar orbit. Then you retrofire again to drop into Earth's gravity well.

Should we ever want to get to the inner planets (and certainly we've already sent a lot of unmanned probes there, and hopefully will continue to do so), we'd do the two stages in reverse order: get to the edge of Earth's gravity well, retro out of the Earth-solar orbit and drop down to Venus or Mercury, retro-burn and drop into the planet's gravitational well. Then to come back, we do a posigrade (forward) burn to escape the planet and move outward from the Sun, meet the Earth and do a retro-burn to enter its gravitational field.

Additional complexities are involved in the fact that the planetary orbits aren't perfectly circular, nor are they perfectly coplanar. There's a need for some course corrections and inclination changes and such along the way. These days, though, after a lot of flights to and from the Intenational Space Station, we pretty much know how to do that.

There are other types of transfers that use even less energy than a Hohmann transfer, but they require much, much longer travel times. Longer travel times means more supplies for the crew, especially consumables. It's likely that a Hohmann transfer will be used for a manned trip to Mars, in mostly the same way Apollo went to the Moon. By going back to the Moon we can practice these very complicated orbital dynamics and become proficient at them.

CHAPTER II:
THE DINOSAURS DIDN' T
HAVE A SPACE PROGRAM—
WE DO

I have yet to see a Hollywood movie that was even close to what a real comet impact on Earth might be like. Hollywood seems to think that we have all these secret spacecraft ready and waiting to take a bunch of nukes up to a comet and just blow it up. Well, where do you even start with that kind of silliness?

Think of the largest nuclear weapon man has ever made. This is the Russian Tsar Bomb. That thing was near fifty megatons of explosive power and had a total destruct radius of ten to twenty miles or so. I guess this is the data Hollywood writers use to decide that the Tsar Bomb could destroy a comet twenty miles in radius. That is just plain goofy. The nuclear device causes so much devastation on Earth because Earth has an atmosphere. A huge pressure wave is created that goes off in all directions from the blast. That pressure wave is what blows houses and trees and everything else down.

We've actually dug holes in the Earth before and set off nukes in them. The blast cracks the dirt and then fuses it all together from the heat. Very little damage is actually done. Well, if you don't count the radiation. It is likely that a nuke would do very little to a comet of any appreciable size. Much more energy would need to be imparted to it. And besides that, we don't have any giant nukes in existence any longer. We've destroyed them all.

This led me to the idea for finding a magic bullet that would destroy a comet. I wanted to do a scaled experiment but we need something that would scale as a comet. I read up on all the latest comet space mission data and realized from looking at drawings of comets and reading about them that they looked a lot like a watermelon. If you used a frozen watermelon then you had a very good approximation of a comet. A comet has a hard crust exterior with a softer interior with pockets of empty space spread about the inside. Hmm . . . just like a watermelon.

When I told the guys I wanted to blow up and shoot watermelons, believe me, it took very little negotiation to get them on board with the plan. After we shot several of them off of a stump we thought it just wasn't sporty enough. Rog had the idea for us to build a trebuchet and launch them. I actually liked that idea for three reasons:

1. Building a trebuchet sounded like fun
2. Shooting the melons in the air and watching the debris field fall might tell us more about how effective our comet defense approach would be
And 3. I mean, come on, we were launching watermelons and shooting them out of the sky!

After a while of shooting the things we realized that we needed a certain kind of bullet instead of shotguns. A .243-caliber rifle bullet with a polymer tip is what it took to destroy the things. I wanted to see what would happen if we hit that thing in the air with the magic bullet so we built a robot. Well, our robot never did do a direct hit on the melons because of a slow data link between our web camera and the computer. We did actually nick one once, but never a direct hit.

This is where Daddy comes in. Now, as long as I can recall, my daddy has been the guy that everybody comes to for hunting advice, gunsmithing, and shooting advice. He has hit prairie dogs at nearly 500 yards and I watched him shoot flying crows with a .22 rifle. So, right then and there on the spot I asked Daddy, "Do you think we could pull that rifle off the robot and you shoot it?"

"I'll try," he said.

"Charlie, I'll bet you can hit it," his friend Mert replied. Mert was sitting in the back of an old pickup watching. After all, we were on Mert's property.

Daddy picked up the rifle and loaded it once we got it off the robot. He held it up to his shoulder for a minute or so and looked down the scope at a few things. Then he told us he was ready to go.

Rog loaded up the trebuchet and when Daddy yelled "pull," he let 'er rip. The melon came flying toward us as Daddy tracked it with the rifle. Honest to God, the instant he pulled that trigger the melon disintegrated into millions of water droplets. He hit it the *first try*! Our jubilation in that scene is genuine because we didn't expect him to hit it the first try. Oh, I knew he would hit it. But I expected him to need some practice. I guess he has had over sixty years of practice and that was enough.

In the end, even though our robot didn't work, we learned that if we needed to scale this experiment up to kill a real comet that we'd need a really big gun and a really big bullet. The bullet would have to be trailer-truck-sized, moving faster than any gun could push it. Perhaps there is future technology that could be developed, like a space-based railgun or something similar that might work. I actually taught a class at the local university where we investigated potential space vehicle designs for comet defense. The only solution the kids came up with was a railgun that had to be built at the equator of the Moon. It wrapped around the Moon completely. The projectile would have to be accelerated around the Moon nearly three times before it would reach the speed needed to destroy a big comet. The class had come up with a very expensive and difficult engineering solution but at least it was a solution. And it all started with a backyard experiment us rednecks did blowin' up watermelons.

★ ★ ★

The extinction of the dinosaurs as we know them occurred at the boundary of the Cretaceous and Tertiary Periods in the geologic timetable. These days, the Tertiary Period is no more, as the International Commission on Stratigraphy has discouraged the use of "tertiary" as a proper name for a unit or rock layer and renamed the period in question, as well as splitting it in two pieces. What once was the Tertiary Period is now called the Paleogene and Neogene Periods, plural. However, the extinction event is still most commonly referred to as the Cretaceous-Tertiary Event, or K-T Event for short. The K-T

Event very definitely marks the boundary between the Mesozoic Era and the Cenozoic Era, in a decided way. Some eighty-five percent of all species on Earth became extinct at this point in time. This includes, as I just mentioned, all dinosaurs as we know them—tyrannosaurs, plesiosaurs, pterosaurs, ceratopsids, all of them—died within a very short space of time, geologically speaking. In addition, many plants and invertebrates died off as well, right down to the microbiological level.

Up until 1980 there were many theories behind the event, none of which quite covered everything. Then Nobel Prize winning physicist Luis Alvarez had a talk with his geologist son Walter over the odd, thin stratum (layer) of rock that almost always occurred right at the boundary, between rocks of the Cretaceous Period and rocks of the Tertiary Period. It didn't matter where in the world you were, if you dug down to those layers of rocks, you'd find this thin layer of a greenish clay. And this thin layer had some unusual properties. They brought in friends, the chemists Frank Asaro and Helen Michel, and together the four studied this layer of rock, noting its peculiarities, not the least of which was an abundance of the element iridium. It is, in fact, the rarest transition metal in the Earth's crust, and it is presumed that most of it wound up sinking with the iron to the Earth's core as the young planet formed.

But not in the boundary layer. There was anywhere from 20 to 160 times the normal amount of iridium in the K-T boundary layer, depending on where in the world you were. Not only that, but the distribution of iridium isotopes was unusual, too. It wasn't the normal distribution you'd find even in iridium ore. No, it was much more like the isotope distribution found in space rocks—meteorites.

And not only did this layer contain too much iridium. It also contained shocked grains of quartz and tiny glass beads that appeared to be tektites.

And suddenly Luis had an idea. Maybe the extinction event had come about through cosmic means. Maybe the dinos died because a really big rock slammed into Earth and wreaked havoc.

Today that is the single widest accepted idea behind the event, but the concept has evolved and enlarged to explain a few more details. The first detail, of course, was to find the "smoking gun"—the proof in the form of an impact formation—an asteroid crater.

Two years before, Glen Penfield, a geophysicist prospecting for a petrochemical company, identified an interesting region that began on the Yucatan Peninsula of Mexico and extended out into the Gulf of Mexico. Oddly enough, the samples they found in the area contained shocked quartz and tektites, as well as igneous and metamorphic rock—including large quantities of andesitic glass—that was not common to the region. It took twelve years before anyone but Penfield and his partner Antonio Camargo recognized it for what it was—a truly huge impact crater. And of just the size the Alvarez team was looking for.

The Chixulub crater, named for a village near the center of the structure, is slightly oval, indicating a somewhat angled descent for the asteroid, and approximately 110 miles (180 km) across. Subsequent studies indicate that this is just the central portion of the crater and the full impact structure is 190 miles (300 km) across. It is one of the largest impact features on Earth. It is also about sixty-five million years old, almost the perfect age for the extinction, though it is estimated to be perhaps 300,000 years too early, give or take. The asteroid itself would have been a little over six miles (10 km) wide and would have struck with a kinetic energy equivalent to about ninety-six teratons (96 million megatons; 4×10^{23} J) of TNT. By way of comparison, the largest nuclear weapon ever detonated on Earth was the Soviet Union's Tsar Bomba test, at only fifty megatons (2.1×10^{17} J). And the volcanic eruption that created the La Garita Caldera—a titanic eruption that devastated the entire region of what is now Colorado state some 28 to 26 million years ago, and ejected enough material to fill Lake Michigan—in the San Juan Mountains of Colorado was only 240 gigatons (240,000 megatons; 10^{21} J).

At the time, the area was even more water-covered than it is now. It would have generated a mega-tsunami that at its peak would have been nearly two miles (3 km) high. The debris from this mega-tsunami has been found in *Missouri*! The northern Gulf coast was, apparently, not too much different in extent than today, and that tsunami would still be over a mile high when it reached the area where the Rocket City would be built millions of years later. Imagine a wall of water bearing down on you that was over a mile up. That's what the dinosaurs around here got to see. The last thing they ever got to see.

In under a second, the entire asteroid would have dug its way under the surface, throwing up superheated steam and molten rock, including not only part of the asteroid but a significant part of the material excavated from the crater. Some of this ejecta would have gone exoatmospheric. Some of it might have even reached escape velocity and be floating out in space somewhere—we are pretty certain rocks from Mars made it to Earth in a similar way, so rocks from Earth might likewise have made it to Mars or elsewhere in the solar system. The rest of it rained back down world-wide, remelting on reentry, and igniting wildfires wherever it landed. The titanic shock waves would have triggered world-wide earthquakes and volcanic eruptions, as well as better than hurricane-force winds.

Carbon dioxide would have been released from the local carbonate rocks, resulting in a short-lived greenhouse effect, a world-wide layer of ash would have been laid down (hence the nice layer of greenish clay at the K-T boundary), then acid rain would begin to fall as dust and ash in the upper layers of the atmosphere dimmed the Sun's light and cooled down the planet, washing the carbon dioxide and sulfuric components from the air. Temperatures would plummet for a considerable length of time. And this is only from the Chixulub impact. Like I said, the current theory expands on that.

You see, the asteroid that hit the Yucatan is believed to have been one of a family of asteroids produced during a collision in the asteroid belt, either the Baptistina family or the Flora family. The parent asteroid may have been as large as 110 miles (170 km) across. Amazing what chemistry and spectroscopy can do.

But this leads us to one of the expansions of the idea. What if Chixulub wasn't the only impact event at the time? What if this was like another Shoemaker-Levy 9 event? Well, it turns out that there are at least three other locations on Earth that might fit the bill. And that doesn't even count Luis Alvarez's original idea that Iceland might be a marker for an impact, occurring as it does along the Mid-Atlantic Ridge tectonic plate boundary.

There's Boltysh Crater in the Ukraine. It's fifteen miles (24 km) across, and 1,800 feet (550 m) deep. Its ejecta blanket is estimated to have covered 9,700 square miles (25,000 km²), and been over a yard (meter) thick, increasing to a depth of 2,000 feet (600 m) at the crater rim. It is roughly the same age as Chixulub, within the uncertainties of the measurements, which is about half a million years. (Hey, that's close, in geologic terms.)

Then there's the Silverpit Crater. This one happens to be in the North Sea off the northern coast of Great Britain. It's about two miles (3 km) wide, but concentric rings where the area slumped (which Chixulub also has) indicate the original crater may have been up to 12.5 miles (20 km) wide. It's still not conclusive that this is an impact crater, but it is highly probable. If so, its impactor would have been about 394 feet (120 m) across, had a mass of around 4.4 billion pounds (2×10^9 kg), and hit the Earth at a speed of around 45,000–112,000 mph (20–50 km/s). It would have also created a huge tsunami and shock wave. The error boundary for when it occurred is quite a bit larger than either Chixulub or Boltysh, but is still centered around the K-T event.

The Shiva Crater is a bit more problematic, mostly because there's a big argument going about whether it's really *anything* or not. It is currently hotly disputed, with some arguing for impact crater, others arguing that it is merely a natural depression or series of faults, and still others concluding it's all in the eye of the beholder and isn't really anything at all. As yet the Earth Impact Database hasn't recognized it, but that might change in the future. The presence of high seismic activity, many faults, and tectonic plate boundaries in the area are likely to have distorted the shape and structure of any impact crater in the area.

But let's assume for a minute that it is an impact crater. It happens to lie partly on the Indian continental shelf and partly in the Indian Ocean west of Mumbai—which means another mega-tsunami, because that area is entirely under water and has been for a long time. It's sort of teardrop shaped, but squared off (thought to be the effect of all those faults), and measures about 370 x 250 miles (600 x 400 km), or about 310 miles (500 km) across overall. The teardrop shape is interpreted to mean the object came in at a low angle—I say object because this one may have been an asteroid, or it may have been a comet—but whatever it was, it is estimated to have been about twenty-five miles (40 km) in diameter. It is reported to contain alkaline melt rocks, shocked quartz, and higher than normal concentrations of iridium. And most importantly, it's the right age for the K-T event.

So, that's four candidates for impacts around the K-T extinction.

(Five, if you consider Alvarez' idea that an impactor punched through the plate boundary to create Iceland.) Now remember all of those effects I described for the Chixulub impact: mega-tsunamis, shock waves, molten ejecta, wildfires, instant greenhouse followed by powerful cooling, acid rain, reduced sunlight, ash all over. Multiply this times four or five. Then add in something called marine regression, which is when submerged sea floor stops being submerged, usually because it has been raised up by . . . earthquakes, which have been happening a lot with all these big rocks smacking the Earth. Then add in the volcanos.

Like the Deccan Traps. These, like the Shiva Crater, are associated with India, and may be part of the same event, presuming Shiva is an impact site. The Deccan Traps is a huge volcanic field in India, laying out gobs of what is called flood basalt, because basaltic lava is runny and thin and it ran all over everything for miles and miles. It erupted multiple times in relatively short order. The end result was that over 193,000 square miles (500,000 km²) of land was covered with lava 6,526 feet (~2 km) deep. That comes to a volume of lava of 123,000 cubic miles (512,000 km³).

Lots of big rocks smacking down.

Mega-tsunamis.

Mega-shock waves.

Global ash.

Flaming ejecta.

Wildfires.

Super-hurricane winds.

Mammoth earthquakes.

Titanic volcanic eruptions flooding lava everywhere.

Sudden warming, followed by rapid, deep cooling.

Acid rain.

Dust and ash blotting out the Sun.

Now imagine that happening all over the world. At more or less the same time.

And that's what happened to the dinosaurs. They didn't stand a chance.

Okay, you say. It was bad for the dinosaurs. But that's ancient history—literally. We don't have anything to worry about.

Wrong.

Ever heard of the Tunguska Event?

On June 30, 1908, a little past 7 A.M. local, waaay up in Siberia, near a river called Podkamennaya Tunguska, something . . . big . . . happened. Something fell from the sky and created devastation. Devastation for miles and miles that could be seen and heard around the world.

Surprisingly for Siberia, a lot of people saw, heard, felt it, but what they observed seems to have depended on how far away from it they were. And it was observable a long, long way away.

Whatever it was, it exploded in mid-air with a force somewhere between five and thirty megatons (21–130 PJ) of TNT, probably closer to 10 to 15 megatons (42–63 PJ). That's more than 1000 times the power of the Hiroshima bomb, it's the equivalent to the US' Castle Bravo test in 1954, and about one third the energy of the Tsar Bomba (you remember that one—the biggest nuke humankind has ever detonated). It created a shock wave that registered on seismic stations across Asia and Europe with a Richter magnitude of 5.0. The atmospheric pressure wave was detected in Great Britain.

Eurasian night skies glowed, and the Smithsonian Astrophysical Observatory and Mount Wilson Observatory both recorded reduced atmospheric transparency and poor seeing for several months afterward. Whatever blew put a buncha crap way up high in the atmosphere.

And that was with the thing exploding 3 to 6 miles (5–10 km) up in the air.

Due to the remote location, little investigation was done for several decades, the first of many expeditions taking place in 1921, and even that by accident, in the course of a survey. Expeditions continued, and the area is still closely studied to this day. Eyewitness accounts, taken both at the time of the event and during later expeditions, were interesting, to say the least.

★ ★ ★

Siberian Life newspaper, July 27, 1908:

"When the meteorite fell, strong tremors in the ground were observed, and near the Lovat village of the Kansk uezd [principality, county] two strong explosions were heard, as if from large-caliber artillery."

★ ★ ★

Sibir newspaper, July 2, 1908:

"On the 17th of June, around 9 A.M. in the morning, we observed an unusual natural occurrence. In the north Karelinski village [about 130 miles north of Kirensk] the peasants saw to the northwest, rather high above the horizon, some strangely bright (impossible to look at) bluish-white heavenly body, which for ten minutes moved downwards. The body appeared as a "pipe," i.e. a cylinder. The sky was cloudless; only a small dark cloud was observed in the general direction of the bright body. It was hot and dry. As the body neared the ground (forest), the bright body seemed to smudge, and then turned into a giant billow of black smoke, and a loud knocking (not thunder) was heard, as if large stones were falling, or artillery was fired. All buildings shook. At the same time the cloud began emitting flames of uncertain shapes. All villagers were stricken with panic and took to the streets; women cried, thinking it was the end of the world. The author of these lines was meantime in the forest about six verst (about four miles) north of Kirensk, and heard to the northeast some kind of artillery barrage, that repeated in intervals of fifteen minutes at least ten times. In Kirensk in a few buildings in the walls facing north east window glass shook."

★ ★ ★

Testimony of S. Semenov,
as recorded by Leonid Kulik's expedition in 1930:

"At breakfast time I was sitting by the house at Vanavara Trading Post [sixty-five kilometres/forty miles south of the explosion], facing north. I suddenly saw that directly to the north, over Onkoul's Tunguska Road, the sky split in two and fire appeared high and wide over the forest [as Semenov showed, about fifty degrees up—expedition note]. The split in the sky grew larger, and the entire northern side was covered with fire. At that moment I became so hot that I couldn't bear it, as if my shirt was on fire; from the northern side, where the fire was, came strong heat. I wanted to tear off my shirt and throw it down, but then the sky shut closed, and a strong thump sounded, and I was thrown a few metres. I lost my senses for a moment, but then my wife ran out and led me to the house. After that such noise came, as if rocks were falling or cannons were firing, the earth shook, and when I was on the ground, I pressed my head down, fearing rocks would smash it. When the sky opened up, hot wind raced between the houses, like from cannons, which left traces in the ground like pathways, and it damaged some crops. Later we saw that many windows were shattered, and in the barn a part of the iron lock snapped."

★ ★ ★

Testimony of Chuchan of Shanyagir tribe, as recorded by I.M. Suslov in 1926:

"We had a hut by the river with my brother Chekaren. We were sleeping. Suddenly we both woke up at the same time. Somebody shoved us. We heard whistling and felt strong wind. Chekaren said, 'Can you hear all those birds flying overhead?' We were both in the hut, couldn't see what was going on outside. Suddenly, I got shoved again, this time so hard I fell into the fire. I got scared. Chekaren got scared too. We started crying out for father, mother, brother, but no one answered. There was noise beyond the hut, we could hear trees falling down. Chekaren and I got out of our sleeping bags and wanted to run out, but then the thunder struck. This was the first thunder. The earth began to move and rock, wind hit our hut and knocked it over. My body was pushed down by sticks, but my head was in the clear. Then I saw a wonder: trees were falling, the branches were on fire, it became mighty bright, how can I say this, as if there was a second sun, my eyes were hurting, I even closed them. It was like what the Russians call lightning. And immediately there was a loud thunderclap. This was the second thunder. The morning was sunny, there were no clouds, our Sun was shining brightly as usual, and suddenly there came a second one!

"Chekaren and I had some difficulty getting out from under the remains of our hut. Then we saw that above, but in a different place, there was another flash, and loud thunder came. This was the third thunder strike. Wind came again, knocked us off our feet, struck against the fallen trees.

"We looked at the fallen trees, watched the tree tops get snapped off, watched the fires. Suddenly Chekaren yelled, 'Look up' and pointed with his hand. I looked there and saw another flash, and it made another thunder. But the noise was less than before. This was the fourth strike, like normal thunder.

"Now I remember well there was also one more thunder strike, but it was small, and somewhere far away, where the Sun goes to sleep."

Note that with increasing nearness to the event, more explosions were observed. Poor Chuchan and his brother were actually in the edge of the blast zone, and managed to witness the entire event and survive. Given that many animals and other tribespeople did not, they were very lucky.

The first formal expedition to the presumed impact site occurred in 1927, led by Leonid Kulik. They were expecting a large crater, but found something very different.

At ground zero, an area of about five miles (8 km) diameter consisted of nothing but upright, scorched tree trunks, dead and bare, stripped of any branches, let alone foliage. Spreading out around this was a vast area of desolation where the trees had been flattened, more or less radially outward from the center. This area was roughly butterfly-shaped (a pattern known now from aerial nuclear blasts), and some forty-three by thirty-four miles (70 x 55 km). Overall it flattened some eighty million trees over an 830-square-mile (2,150 km^2) area.

Kulik was curious about the many small "pothole bogs," as he called them, suspecting they were small impact craters, but when he, with some effort, drained and cleared one, he found a tree stump at the bottom. This may or may not have been the fate of some of the trees destroyed in the explosion.

The matter was argued for many decades, with a bunch of wild theories being flung about, but it's now believed that an object—a comet, an asteroid, a cometary nucleus—came in from space over Tunguska and exploded. The shock wave propagated down as well as out, which caused the trees directly beneath to be stripped but left standing, and radially flattened those farther out. Ice crystals blown high in the atmosphere produced noctilucent clouds (reflecting and refracting sunlight from below the horizon due to their altitude) over the continent, and spreading some fine debris aloft as well, darkening the sky somewhat.

Since then, the following data have been discovered:

Microscopic spheres of silicate and magnetite were found

embedded in the trees and soils of the affected area. These spheres contained high proportions of nickel relative to iron, a property common to nickel-iron meteorites but not to terrestrial magnetite.

Peat bogs in the region show considerable isotopic element ratio inconsistencies in the layer for 1908, but not before or after. This is certainly evidence that something unusual occurred there and that it was likely an impactor from outer space. But the biggest key data was iridium. High amounts of iridium were found in the area, not unlike that in the K-T boundary layer.

In more recent years, actual impactors have been located, if not actually found, since they are buried deep. They have been located via such technologies as ground-penetrating radar, seismograms, and magnetograms. In 2010, an ice impactor was found in a crater buried under modern permafrost. And nearby Lake Cheko has proven to be a very recent addition to the bodies of water in the region. The lake is only about 100 years old. Its sediments show absolutely no indication of plant life older than 1908, but an abundance of it after. It has a conical bed shape, much like an impact crater would produce. The lake is elongated, some 2,323 x 1,194 feet (708 x 364 m), with the long axis pointing to the hypocenter of the explosion, about 4.3 miles (7 km) away. And most telling of all, magnetic readings indicate there may be a large, three-foot (~1 m) rock under the lake's deepest point.

The Tunguska event may be a century old, but it isn't ancient history. And, we still do not fully understand all that happened there. How many impacts? An asteroid? Comet? Cometary nucleus? Why did it burst in the air? There are millions of questions to be answered. But the biggest question is, "How do we protect ourselves from such impacts?"

★ ★ ★

Still not satisfied?

July 1994.
Jupiter.
Comet Shoemaker-Levy 9. The "String of Pearls" comet.

Shoemaker-Levy 9 was discovered by no less than Eugene and Carolyn Shoemaker, and their partner David Levy, in early 1993. (Sadly, Gene, a geologist turned astronomer after helping to found the field of astrogeology from years of studying impact craters on

Earth, is no longer with us. He died in a head-on car crash near Alice Springs, Northern Territory, Australia in 1997 while on an impact crater hunt.) It was the first comet found orbiting a planet—Jupiter— probably captured by that planet a few decades earlier, and in a highly eccentric (elongated) orbit, with an orbital period of around two years, give or take. During its previous pass by Jupiter in about 1992, it came inside Jupiter's Roche lobe and was gravitationally torn apart, but the pieces remained grouped together.

By 1994 it was approaching perijove (closest approach to Jupiter) again—only this time, perijove was *inside* Jupiter; it was going to hit. And in July of that year, the world watched in awe tinged with some horror as pieces of Shoemaker-Levy 9, ranging from small chunks up to 1.2-mile (2 km) fragments, struck Jupiter's southern hemisphere at speeds of thirty-seven miles per second (over 134,000 mph or 60 km/s). An entire fleet of unmanned spacecraft, including Hubble Space Telescope, ROSAT X-ray satellite, the Galileo probe already en route to Jupiter, and even the Ulysses solar observer and Voyager 2 (past its rendezvous with Neptune and headed out of the solar system) were trained on Jupiter, ready to watch the "big event."

The first impactor, fragment A, was a moderate sized piece of the comet. It still produced a mushroom cloud whose temperature reached 43,200°F (24,000K), well more than 42,000°F (23,500K) hotter than Jupiter's cloud tops. The mushroom cloud reached a height of 1,864 miles (3,000 km) before finally slumping back down and spreading out over the Jovian cloud tops, turning dark. The resulting "scar" on the face of Jupiter was . . . one Earth radius across.

And that was just the first, medium sized impact.

During the course of a week, chunk after chunk of the comet bombarded Jupiter, creating huge dark crescent-shaped marks (presumed to be organics formed by chemical reactions in Jupiter's upper atmosphere, possibly including reactions with cometary material). These marks were more easily visible than the Red Spot, even in small amateur telescopes, and appeared to be oriented in the direction of angle of impact. They sent seismic waves (earthquake waves) flying through Jupiter at speeds of 280 miles per second (1,007,000 mph or 450 km/s) and continued for over two hours, back and forth from impact to antipode (straight through the planetary center to the opposite side) and back. They created aurorae at the impact sites and at the site antipodes.

The largest piece, fragment G (they were assigned letters in alphabetical order of arrival), 1.2 miles (2 km) in diameter, struck two days after the first impact. It produced a titanic fireball, which settled down into a dark "scar" more than 7,456 miles (12,000 km) in diameter—bigger than the Earth—and released kinetic energy equivalent to six petatons (6,000,000 megatons, or 25 million PJ) of TNT. If you piled all of the world's nuclear arsenals into one big heap and detonated them all at once, you would need to detonate 599 more piles just like it to equal the energy released in that single fragment G impact.

Two more impacts the next day were roughly equivalent to fragment G.

Impacts continued until July 22, when fragment W impacted Jupiter.

Imagine if even one of them had hit Earth.

In addition, an investigation of Jupiter's moons has found no less than thirteen chains of craters on Callisto, and three more on Ganymede. Evidently this sort of event isn't uncommon.

And in July 2009, fifteen years later, another dark spot showed up on Jupiter, fully the size of the Pacific Ocean. Evidently another body had impacted Jupiter, although since no one was looking for it, it wasn't seen. At the time of its discovery it was still warm, and spectroscopic studies showed hot ammonia and dust in the upper levels of Jupiter's atmosphere in the region of the "scar." This, then, was something with a little more heft than a comet, likely an asteroid.

So impacts keep happening. In recent decades astronomers and even casual observers have been witness to near misses by some fairly good sized space rocks. Some have even been videotaped. Astronomers have thus discovered an entirely new class of asteroids, separate from the well-known asteroid belt between Mars and Jupiter, that they have dubbed Near Earth Objects, or NEOs. And the danger to Earth is from these NEOs.

Think of the solar system as like a pool table. The Sun and planets are the pockets, and the asteroids and comets are the pool balls. Only the pool balls and the pockets are continually moving, and the pool table isn't flat—every pocket has a depression around it. The bigger the pocket, the bigger the depression in the pool table. So when a pool ball comes near a pocket, even though it might not go in the pocket, its direction of motion might be affected by the depression around the pocket. So the movements of the pool balls are constantly changing. And if a given pool ball keeps bouncing around the rim of the table and coming back by the same pocket time after time, sooner or later, it's liable to fall in.

That's Earth's situation with the NEOs.

When the astronomers finally presented the government with enough evidence that there were scads of these billiard balls out there, whizzing by Earth's corner pocket, the government decided it was time to do something. And since NASA was the space-based government agency, NASA got the directive to do something. As usual with space policy and space politics, the politicians used this as a soundbite and ordered NASA to solve the problem. Of course, the politicians didn't take the risk of giving NASA any extra budget to do the effort. In fact, they didn't get any more money to do it, just got directed to do it.

★ ★ ★

The NASA Authorization Act of 2005 states in part:

The U.S. Congress has declared that the general welfare and security of the United States require that the unique competence of NASA be directed to detecting, tracking, cataloguing, and characterizing near-Earth asteroids and comets in order to provide warning and mitigation of the potential hazard of such near-Earth objects to the Earth. The NASA Administrator shall plan, develop, and implement a Near-Earth Object Survey program to detect, track, catalogue, and characterize the physical characteristics of near-Earth objects equal to or greater than 140 meters in diameter in order to assess the threat of such near-Earth objects to the Earth. It shall be the goal of the Survey program to achieve 90% completion of its near-Earth object catalogue (based on statistically predicted populations of near-Earth objects) within 15 years after the date of enactment of this Act. The NASA Administrator shall transmit to Congress not later than 1 year after the date of enactment of this Act an initial report that provides the following: (A) An analysis of possible alternatives that NASA may employ to carry out the Survey program, including ground-based and space-based alternatives with technical descriptions. (B) A recommended option and proposed budget to carry out the Survey program pursuant to the recommended option. (C) Analysis of possible alternatives that NASA could employ to divert an object on a likely collision course with Earth.

The result of this directive was a report presented to Congress in early March 2007. This was an Analysis of Alternatives (AoA) study led by NASA's Program Analysis and Evaluation (PA&E) office with support from outside consultants, the Aerospace Corporation, NASA Langley Research Center (LaRC), and SAIC (amongst others).

★ ★ ★

Now, it turns out that the U.S. government wasn't the only one to sit up and take notice. ESA and several other space agencies have banded together into what is known as Spaceguard, a worldwide effort to find NEOs before they find us, as it were.

But from the inception of the project until 2008, only 982 had been found. Estimates say they have about eighty percent of them, with about thirty percent of the sky still left to survey. I hope they're right. I, personally, think that their estimates are quite optimistic.

Anyway, the NEOs are further subdivided into PHOs— Potentially Hazardous Objects. These are defined as NEOs that come closer than 0.05 AU (4.6 million miles or 7.5 million km) and have an absolute magnitude of 22.0 or brighter (which is a reasonable indication of large size, assuming an albedo (measure of reflectiveness) of thirteen percent). That assumed reflectivity means the object would be typically 500 feet (150 m) in diameter, and hence capable of wreaking a bit of havoc on Earth. Smaller objects could, without any doubt, cause some problems, and maybe even wipe out a city if they hit in the right places, but would not be in danger of being planet-killers. And planet-killers—the deal breakers—are what Spaceguard is worried about right now.

These can be further broken down into subcategories. The most dangerous of the dangerous are the Earth-crossers. This means that their orbits actually intersect Earth's orbit: if Earth and the asteroid were to arrive at the intersection point at the same time, there would be a collision—an impact. The other interesting subcategory consists of those that are moving in more or less the same general direction as Earth, and so the relative speeds are small. We—Earth as a whole—have actually mounted two lander missions to asteroids fitting this latter description. JAXA's Hyabusa probe was sent to 25143 Itokawa, and NEAR-Shoemaker was sent by NASA to asteroid 433 Eros.

In general, however, NEOs are divided into three families: those whose orbits are inside Earth's orbit—the Atens; those whose orbits cross Earth's orbit—the Apollos; and those whose orbits are largely outside Earth's orbit—the Amors. For those asteroids whose orbit is very Earthlike, there is also the Arjuna category, which can overlap any of the other three.

But remember, just because it isn't an Apollo asteroid now doesn't mean it won't be in the future. Amors often cross Mars' orbit, and a close encounter with it could change the asteroid's orbit and send it hurtling in—or out—of its current orbit. Likewise with Atens and Venus (and occasionally Mercury or the sun).

And don't forget comets. By 2010, we knew of 84 NEO comets, too. And every close approach to the Sun results in outgassing, which is effectively a small rocket, changing the orbit a little bit each time.

Cosmic billiards, where the stakes are, at the least, the survival of the human race.

★ ★ ★

Known Impacts in the Twentieth and Early Twentiy-first Centuries

Year	Event	Details
1908	Tunguska Event	10 megaton airburst over Siberia; asteroid/comet/comet nucleus

Year	Event	Details
1979	Vela Incident	Spy satellite detects an event between south Atlantic & Indian Oceans; nuclear or non-nuclear nature undetermined.
2002	Eastern Mediterranean Event	33 feet (10 m) diameter object airburst over Mediterranean Sea; 26 kiloton yield.
2008	Sudan Event	2008 TC3 detected; predicted to impact Sudan the next day; time given. First accurate prediction of an impact event.
2009	Indonesia Event	Fireball yield of 50 kilotons, believed to be 33 ft (10 m) diameter.

Near Misses in the Twentieth and Early Twenty-first Centuries

Year	Event	Description
1972	The Great Daylight Fireball	Observed over Rocky Mountains from Utah, USA, to Alberta, Canada. Filmed in the Grand Tetons of Wyoming; recorded by satellites. 9.8 ft (3 m) dia; atm graze; closest approach to surface 35 mi (57 km).

Year	Event	Description
1989	4581 Aesclepius	This Apollo asteroid, 1,000 ft (300 m) wide, passed within 430,000 mi (700,000 km) of Earth. If it had been 6 hrs earlier it would have impacted with a force of 12 times Tsar Bomba.
2004	2004FH	LINEAR determines 100 ft (30 m) asteroid will pass Earth at 26,500 mi (42,600 km), 1/10 distance to Moon.
2004	2004FU162	2 wks after 2004FH, 20 ft (6 m) asteroid passes 4,000 mi (6,500 km) away from Earth, 1/60 distance to Moon, and just over 1 Earth radius from the surface. Only detected hours before closest approach. Likely to have burned up in atm.
2009	2009DD45	Flew within 45,000 mi (72,000 km) of Earth, 2x height of geostationary satellites. 115 ft (35 m) in diameter.
2010	2010AL30	Closest approach 76,000 mi (122,000 km). Size 33-49 ft (10-15 m). If it had struck, energy yield would have been 50–100 kilotons. (Hiroshima bomb only 13–18 kilotons.)

Year	Event	Description
2011	2011MD	Closest approach 12,000 mi (20,000 km), or one Earth diameter. Size 16-66 ft (5-20 m). Passed over Atlantic Ocean.
2011	2005YU55	Closest approach 201,700 mi (324,600 km); 85% of distance to Moon. Size 1,333 ft (400 m).
2012	2012BX34	10 weeks after 2005YU55 passby. Closest approach 37,000 mi (60,000 km); about 3 Earth diameters. Size 37 ft (11 m).
2012	2012 DA14	Closest approach Feb 16, 1.5 million miles (2,500,000 km). Size 150 ft (45.7 m) diameter.

Future Predictions of Near-Misses and/or Impacts

Year	Event	Description
2013	2012 DA14	Close pass estimated 16,700 mi (27,000 km; 6.25 Earth diameters; inside geostationary orbit). Size 150 ft (45.7 m) diameter. Probability of impact 1 in 77,000. If impactor, Tunguska sized event.
2019	(89959) 2002 NT7	Close pass, 0.4078 AU (37,910,000 mi; 61,010,000 km). Size, 1.2 mi (2 km).

Year	Event	Description
2029	99942 Apophis	Close pass; possible "keyhole pass" that would align it for a collision in 2036. Probability of keyhole pass low. Size, 885 ft (270 m).
2032	(144898) 2004 VD17	Close pass, 0.02 AU (1,900,000 mi; 3,000,000 km) from Earth. Anticipate refining orbit at this pass. Crosses orbits of Venus, Earth, Mars. Lots of opportunity for orbit interaction.
2036	99942 Apophis	Possible "keyhole pass" collision
2102	(144898) 2004 VD17	Possible impact; currently deemed unlikely. Awaiting refinement of orbit from observations in 2032. Size (580 m). Impact would generate 6-mi (10 km) crater, 7.4 magnitude quake.
2880	(29075) 1950 DA	Size 0.6 mi (1 km). Probability of impact 1 in 300.

★ ★ ★

Note that a six-mile (10 km) diameter asteroid is considered an extinction-level event, on a par with the K-T Event. Also note that, if 2012 DA14 were actually going to strike the Earth in 2013, it is already too late as of this writing to do anything about it, with *any* technique. And it had already crossed Earth's path three times before it was ever noticed.

The next question becomes, have we really found most of 'em, or not? Well, the astronomers assigned to the task are working really hard, so let's hope they have. The evidence indicates, however—especially given 2012 DA14—that they probably haven't.

The tricky thing about asteroids is like what the military guys say—it's the one you *don't* see that gets you in the end. That's because if it's coming straight at you, there isn't any sideways motion to give it away, and it appears like an ordinary star—until it gets close enough to look bigger and brighter. But by then we would be in deep trouble. That's why we're trying to find them while they're still a ways out, and why the ones that barely zipped by us, we had only hours' warning about. Astronomers also have to have enough observations of the thing to be able to calculate an orbit, which can get tricky in itself. But note, in the future predictions table, that the last three entries have some calculatable probability of an Earth collision, and it wouldn't be pretty.

Even if we did find them all now we aren't in much better shape than the dinosaurs because we can't do anything about them. We would just know our doom was pending with no plan or capability to stop it. In that regard, you might even say the dinosaurs were better off than we are. At least they didn't know the end was coming. Seriously, what do we do, now that we've found 'em?

What do we do? We somehow have to stop them from impacting. But we do not know how to do that. That kind of technology doesn't exist today and won't for many years to come. What do we need to stop an impactor once we've determined it's on a collision course?

That depends on what we have available. And what we have available right now is essentially nothing.

Yes, that's right. Zip. Zilch. Nada. The Big Ø. Because you can't just stop an asteroid or comet. The momentum is too great. If you tried—assuming you had the capability—it would likely shatter, and that's just as bad as if you hadn't done anything, because now you have the same mass coming at you like a shotgun blast. Same thing goes for trying to blow it up. It didn't really help Jupiter that Comet Shoemaker-Levy 9 was in pieces instead of one big chunk. It just distributed the energy over that much larger an area.

In other words, "speak softly" scores bigger points than "big stick." Unless that big stick is so large that it totally vaporizes the impending object of doom.

So . . . where does *that* leave us?

With zip. Zilch. Nada. We are, in essence, for real, literally, up the creek with no paddle in sight.

We don't currently have a decent way to get something heavy out of low Earth orbit, let alone to an asteroid that's truckin' at us at, oh, say, 64,000 mph (~29 km/s), which happens to be the average speed of asteroid Asclepius, a member of the Apollo group of Earth-crossers. Yeah, there's the old science fiction standby of shooting ICBMs equipped with nukes at 'em. But one, ICBMs are designed for suborbital hops, not LEO, and certainly not a transition trajectory. And two, you don't wanna blow the thing up, like I said before. Especially not with a nuke, because then not only do you have a cosmic shotgun blast coming at you, you've got a *radioactive* cosmic shotgun blast coming at you. In fact, nukes aren't big enough anyway. One of the experiments we did on *Rocket City Rednecks* was to shoot different bullets at frozen watermelons to determine how to destroy them. Frozen watermelons are a lot like comets, and it took a bullet that, when scaled up to real life size, would have taken tens to hundreds of Tsar Bombas to destroy the comet. Off camera, we also shot at rocks to simulate asteroids. The situation with the asteroids was even more difficult.

The one possible NEO we could go after (assuming we could get there) is a "rubble pile" asteroid—an asteroid that is very loosely bound together by gravity and maybe a little something that passes for dirt or a gravel pile. Burrowing a nuke, again, a very, very big nuke, into the middle of this and letting loose sends the rubble pile in all directions, with very few fragments aimed at Earth. For a solid object, though, that's pretty much impossible without way more advance planning than we really expect to have.

No, provided we have enough time, meaning many years to decades of knowledge of the coming impact, then the best thing to do is to plan enough in advance to be able to construct what you need, have some decent heavy lift vehicles available, and then use one of a few known techniques to gently nudge the big rock or ice cube hurtling at us into a different orbit. We've already talked about heavy lift vehicles. Now let's talk about nudging.

It would seem that the simplest thing to do is to just launch a nice, massive hunk of metal at the asteroid and let it hit the asteroid at the right angle to shove it the way we want it to go. This has the advantage of being quick and easy and relatively inexpensive, but the disadvantage of creating a shotgun blast if we happen to have misjudged the degree of structural integrity of the asteroid. Besides all that, how do we send this big chunk of metal? Rockets would likely not accelerate the bullet fast enough to make much of an impact on the object. There is a technology gap here in how to make the comet/asteroid killing bullet go fast enough to do the right damage. Also, how accurate of a shot could we really make? At the vast distances in space, being off target by very tiny angles could be huge misses at the target distance. This is called pointing jitter. To understand it, go point a laser pointer at a spot on the wall across the room. Then point the same pointer down your driveway at your mailbox or something farther away. You'll learn how much harder it is to hold the laser on the mailbox because it is so far away.

Another simple thing is to take up several nukes on the heavy lift vehicle, and detonate them, sequentially, *next to* the asteroid, using the force of each blast to give the asteroid a shove. This has the disadvantage of having a whole bunch of nuclear bombs get spewed all over if something should go wrong during launch or ascent. In addition, depending on the distance at which the intercept is forced to take place, there could be significant electromagnetic pulse effects on the ground. For this method, farther away is definitely better than close, and the farther, the better. The problem with this idea is that we would have to know about the object a long time in advance. Rockets are relatively slow when it comes to comets and asteroids because they are so far away. If we see this comet out past Mars how long would it take us to get to it? We have already discussed how it would take almost a year to get to Mars from Earth with modern rockets.

Attaching a rocket engine to the asteroid might work, too, at least as long as the fuel lasted and we, again, knew about the problem with plenty of advanced warning. Possibly the final stage of the heavy lift vehicle could be modified to either land on the asteroid's surface, or javelin into it, then ignite. A variant on this idea is an ion thruster. These use electrostatic or electromagnetic force to accelerate ions out of the engine and have been utilized in a number of solar system probes. Or, you can turn the thing around, and instead of having to mount it on the asteroid, use the ion engine as a kind of "shepherd," letting the ion thrust beam strike the asteroid and push it along. This is called an Ion Beam Shepherd, or IBS. The critical issue with this method is that the IBS must fly in close formation with the asteroid as it changes orbit, due to the beam spread with increasing distance from the engine.

One possible way to nudge the thing is to let the Sun do the work for us. Rather, the light from the Sun. We launch a lander at the asteroid, and on its arrival it unfurls into a solar sail. The solar sail uses the pressure from the Sun's light (yes, light does exert pressure, just a tiny amount so solar sails must be large) and gradually steers the asteroid into a different orbit. Like wind sails, solar sails can be tacked, so that the direction is vectored, and we can then steer the asteroid almost anywhere we want it. The first science fiction short story I wrote and sold was about this idea—using solar sails to steer asteroids is possible but a very large endeavor. Solar sailing also requires that we know the object is coming years and maybe decades in advance.

A solar furnace would be a variant on a solar sail. In this case, the lander unfurls into a hemispherical or preferably parabolic reflector (which could be made of much the same materials as a solar sail), directing and concentrating the sunlight onto a single spot on the asteroid. The material at that spot vaporizes and creates a tiny jet of escaping material, in essence a tiny rocket, which if placed in the right spot gradually alters the orbit of the asteroid. This concept would be more effective on a cometary body, due to its makeup of volatile ices.

The downside to both of these methods is that they depend on the proper unfurling of the mechanism, and on thin sheets of metalized plastic polymer materials, which can tear easily. In addition, like the IBS, the solar furnace would have to be flown in formation with the asteroid or comet.

Or, instead of a solar furnace, we could swap out for a laser beam. We could theoretically build a very large diode laser array that is run off of a nuclear reactor. The laser beams would be focused onto parts of the object to melt it away in jets of vapor. This is another very long, slow process.

If we could do all these other ideas we could just about as likely fly a team of astronauts up there with picks and shovels and other implements of destruction. Not likely and we really have no idea how we could do something like that. There are no mobile oil well drilling Moon Buggies that we could fly up to the asteroid on a Space Shuttle like they did in Hollywood movies.

All of these concepts have been essentially in the current realm in terms of existing technology and techniques. While I wouldn't say they would be easy to do, I would say they are doable. The effectiveness of the above-mentioned approaches would also vary greatly based on the amount of warning we get before the object's approach.

There are many other ideas that range from clever to wild but none of them are immediately available technologies. The majority of the ideas do involve the same general idea—attaching a pusher to the asteroid or comet. One possibility would be to use an unconventional rocket, like a nuclear thermal, or a magnetoplasma rocket. The former has been developed. The NERVA, or Nuclear Engine for Rocket Vehicle Application, was a joint venture between NASA and the Atomic Energy Commission, and it did everything *but* fly, and met or exceeded expecations. The latter was conceived and is being developed by former astronaut Dr. Franklin Chang-Diaz.

His version is called VASIMR, for VAriable Specific Impulse Magnetoplasma Rocket. It heats the fuel with radio or microwaves, then uses an electromagnetic system to accelerate it out of the engine. Because the heating and acceleration system can be controlled, the engine can be throttled up or down over a wide range of thrusts.

The U.S. Navy recently announced in the news that it has a successful railgun prototype weapon. A railgun, or mass driver, is a device that uses two parallel conducting rails and either a conductive projectile or a conductive sabot to electromagnetically accelerate the projectile along the rails to what can be very high speeds. They are just now becoming useful as weapons, but they can be effective for space-based activities, too. Due to the problems with atmospheric friction, Earth to orbit operation may not be practical—but exoatmospheric, there's nothing to stop them!

The railgun uses extremely high voltage and very large magnetic fields to accelerate a projectile much in the same way that a particle accelerator (sometimes called an atom smasher or supercollider) accelerates subatomic particles. The problem with railguns is a purely technical one. The technology is not quite mature and it has never even been really investigated at the large size and power needed for planetary defense and it most certainly has never been designed for being fired in space.

If we figure out a lot of engineering details is it possible that we could use railguns for stopping or pushing comets or asteroids? Well, if you can accelerate a seven pound (3.4 kg) projectile to a speed of 5,400 mph (2.4 km/s), which we can, then that projectile will deliver, on impact, 4.1 million pounds (18.432 million N) of kinetic energy to whatever it hits. The biggest problem right now is that the power source is huge, like room-sized huge. The state of the art railgun made by the Navy is big enough to shoot at large ships over the horizon by tens of miles and maybe farther, but this is several orders of magnitude too small in scale to save Earth from an impending impact from space.

In the episode of *Rocket City Rednecks* where we investigated comet impacts we discovered that the energy required to totally destroy a comet would require accelerating a 10,000 kg mass to 0.3% of the speed of light. That would take a lot of energy, actually, many, many full sized nuclear fission reactors. Also, the projectile would have to be fired in space or the atmosphere would destroy the projectile. The best place nearby to build a planetary defense railgun would be on the Moon. Heck, put several on the Moon, aimed in different directions! You can use one power source for all of them because chances are you'll only need one at a time.

It might even be possible to use a bucket shaped sabot and heave a hunk of Moon rock out there. Same effect, free ammo! In this case we could possibly shoot many times assuming we can recharge the railgun batteries quickly enough. And when we're not using the railgun to heave rocks at threatening asteroids, we can use it as a mass driver to mine the Moon and fling ore, or even smelted raw metal, into one of the Lagrange points for recovery and use on a station outpost or even back on Earth.

Then, when we manage to miniaturize the power supply, we can look at a different notion. We can land a small railgun on the asteroid itself, use some pyro anchors to fasten it down, and start scooping away at the asteroid, putting each scoop in one of those bucket-shaped sabots and heaving it off the asteroid. It won't accelerate as fast, but it doesn't need to. Now you've got near continuous gun recoil affecting the asteroid, and all that kinetic energy becomes thrust. There is still a tremendous amount of technology that would have to be developed. We don't know how to build such a mass driver, and we certainly don't know how to do it in space, on the Moon, or especially something as evasive and unstable as an asteroid or comet. This is why we need to be getting good in space. We need to practice getting there and working there and experimenting with new wild and wacky ideas out there.

Then there's the concept that Gene Shoemaker and Nick Szabo came up with independently, and which Szabo called "cometary aerobraking." The basic idea is to create an atmosphere—however temporary—in front of the object somehow, and create drag on it, causing it to slow down and thus change orbit. This would be simple aerobraking. The "cometary" part comes in if you manage to snag a hunk of cometary ice and put it in the asteroid's path, then vaporize it just as the asteroid gets there. Many means of vaporization could be used, but Szabo suggested just nuking it, direct and simple. Well, maybe. How do we get the nuke there and has anybody really done the fluid dynamics calculations to determine if this really makes sense? Where will the atmosphere really form?

The Yarkovsky effect method is sort of a variation on a solar sail, but in a little bit different way. A portion of the asteroid could be dusted with white titanium dioxide, or alternatively, black soot, in order to change its albedo. Albedo is a measure of the reflectivity of a surface. The more it reflects, the less energy it will absorb, and so the less energy it will radiate away as heat (infrared radiation). Conversely, the more it absorbs, the more it will radiate away as heat. As we've already established, light exerts pressure; it has momentum. Therefore, if we redo the way the asteroid radiates heat, we can create a subtle thrust that will push the asteroid off course. Well, maybe. We still have to get there and deploy the powder in a controllable manner. We have to do this in space. We need to practice being in space.

And then there's the gravitational tractor. It's a slow process, and would probably take several years to accomplish a significant deflection. I'm not a real fan of this idea but who knows. It could possibly work. If it does then it would work whether the object is a comet, a solid asteroid, or a rubble pile. It's really quite simple. Send a very massive unmanned vehicle (essentially a big lump of dense metal, like maybe depleted uranium or tungsten or something similar, using an ion engine, which we don't really have) to the vicinity of the asteroid, and allow the mutual gravitation to attract the asteroid. As the would-be impactor moves toward the tractor, the tractor moves away. This slow, continuous process will eventually drag the object out of orbit and into almost any orbit that we could want, merely by controlling where the tractor goes. The comet and the tractor will dance around each other in a complex two-body gravitationally charged dynamic trajectory.

Ultimately nothing is a sure bet. As long as we remain tied to this "big blue marble" we remain tied to its fate. If a catastrophe befalls it, it befalls us, too. The best way to ensure the survival of the human species is to go by the old saying, "Don't put all your eggs in one basket." We need to establish colonies elsewhere, on planets and satellites in our own solar system, and then move out into the stars. That way, we might lose our home world, but we won't become extinct.

That is the pipe dream or the futurist point of view. We'll colonize the galaxy to make humanity survive forever. Well, sure, some day we may very well do that. But more pragmatically, before that day comes, we need to be prepared to defend our planet, our home, all of humanity as it exists now from a similar fate that fell upon the dinosaurs. The dinosaurs didn't have a space program. We do. If we become extinct because the politicians convinced us to let our space program shrivel and die on the vine then it is our own damned fault. We have a moral imperative to our children and humanity to develop access to all of our solar system with new technologies that will enable us not to go the way of the dinosaurs. But our fate would be worse because we would knowingly be sitting around with our thumbs up our posteriors while philosophizing about the end of man. We can't let politics kill us all!

CHAPTER 12:
THE MILITARY USE
OF SPACE

I want to see Captain Kirk! I want to see the Stargate Command and General O'Neal's team whupping aliens across the galaxy. That is what most of us think of when we say the militarization of space. But in all honesty, it is much more mundane than that for now. As far as I know.

The immediate militarization of space is mostly the use of reconnaissance and communication satellites. We are learning that these things might not be safe if we got into a shooting war with a big nation. This is what led me to the idea of a recon spy balloon.

Our balloon spy platform was designed to do quick recon over trees, buildings, and hills and it was pretty much considered to be disposable. If an enemy saw it they could very simply shoot it down. But it only cost a few hundred bucks so it wouldn't matter.

The really cool part here is that if we had cut the tether on that balloon it would have floated on up to as high as 100,000 feet. That is in an area known as "high altitude" and "near space." It is too high for airplanes but not high enough to be in space. If we could figure out a way to fly the balloon in that region of the atmosphere it is highly unlikely it could be shot down easily. It would take very expensive rockets to shoot it down. Some day in a future season I would really like to build a "near space" vehicle and test it out.

I'd also like to build an actual spy satellite and put it orbit. Now that would be a fun episode. It would have a big budget too! The ride there would be very costly. That is the deal with the militarization of space. It is expensive but, as we have seen in the past, very necessary.

Now if we ever start colonizing space there is where real space warfare will start. War is Hell. Imagine war in an environment that will kill you if you sneeze funny. War in space is even more hellish than terrestrial warfare. It would be far more difficult too. Soldiers would have to all be in spacesuits. Maybe they would need to be in spacesuits with armor.

Building that Junkyard Iron Man was crazy hard. Now I'm thinking about how hard it would be to build it and it also be a spacesuit. New materials and cheap building processes might be the key. Plastic forming techniques like the vacuum technique we used to make the radar absorbing plates for the stealth pickup truck might be one such technology. The base layer of the armored spacesuits should probably be high density polyethylene plastic which can be vacuum formed and shaped easily. It is also fairly bulletproof. Then the plastic pieces would be wrapped in composite fibers like Kevlar or carbon. Actually, a hardsuit constructed of these materials might prove to be a better space suit anyway. Who knows?

One thing I do know is that if we ever are going to do more in space than spy satellites and communications then we will need to be very innovative and imaginative. Maybe we'll come up with a few more ideas we can test out on the show.

★ ★ ★

At some point when we talk about space programs of the world we have to discuss that the elephant is in the living room.

And that elephant, of course, is the militarization of space. For decades it's been perceived that the world governments have agreed not to put the military into space. But this really isn't true. There are spy satellites, there are military communications satellites, and if we ever fire nuclear weapons or intercontinental ballistic missiles, they will most certainly fly through space to get where they're going. So, really and truly, not having any weapons in space may or may not be the truth and/or a bad thing. At some point we're going to have to destroy a malfunctioning satellite in order to protect other spacecraft assets from orbital debris.

We don't really know how to do that yet but we're getting close to the point where we've got to learn how. And in the very near term we might need to put high energy laser systems in orbit that can float around in orbit vaporizing small orbital debris fields. Maybe there are other solutions to that, including transporting large sticky balls that absorb and catch orbital debris, and then the whole thing burns up on reentry. All of these ideas are wild concepts that need a lot of engineering work to determine if they're possible. We need a budget line in our space plan to look into these concepts.

And don't you kid yourselves that other countries in the world are not preparing ways to destroy satellites—the Chinese have already done this. It's quite obvious to most people who live on this planet that we really depend on satellite communications and information gathered by spy satellites. How many times do we use our GPS navigators to find a store or a residence? How often have we used Google maps or Google Earth? Well, civilians use these technologies all the time, and so do the military. It's only sound strategy in the event of war that you would take out the other side's means of communication, intelligence gathering, and tactical operations aids such as navigation. Like I said, don't kid yourself that the nations of the world aren't planning on the militarization of space already.

And how about one day when we finally detect that asteroid, or comet, or near Earth object of some sort that is going to slam into the Earth and cause major problems for us? We will be scrambling to figure out how to put the right type of weapons into space in a very short time frame that may have some impact on our impending doom. It would be a lot simpler if we already had a design for a platform, just completed, prototypes to experiment with, and at least one or two systems deployed in orbit. Perhaps something that's not just in orbit, but on the Moon, or at a Lagrange point or somewhere else in space. But the point is, we need the planetary defense systems, weapons systems, in space now before it's too late and we're invaded by a world killing object such as the one that gave the dinosaurs such a problem.

So we can see that "no weapons in space" really is a bunch of nonsense. And it's not really a good plan. What if we invented a technology that we could deploy in orbit around the Earth that would make nuclear war completely ineffective? Who wouldn't want that to be deployed? What if we *know* that other countries have weapons designed that they will launch from the Earth into space to destroy our satellites? And what if we had a technology that had to be deployed in space, and was a weapon, that could stop this anti-satellite technology? Would we accept that as a weapon with a place in space?

We will always have weapons. We will always need weapons whether we are at war with ourselves or not. There's always the possibility of a natural disaster. And there might even be a billion-to-one, or even larger, or maybe smaller, probability that we could be invaded from outside of Earth.

Yes, I do mean the possibility of an alien invasion. We have no idea if it ever might happen or could happen, but the probability is not zero—especially with the more habitable planets that the Kepler program finds—and maybe the probability of an invasion is not large as far as we know, but it is possible. It makes for great cinema and television, but is it possible? The point of bringing this wild idea of an alien invasion into the mix again is the stressing of the point that if we wait till it happens it'll be too late to place a system into space that will help prevent the disaster. In all seriousness, we should be ashamed of ourselves if we get destroyed by a natural disaster, since we've been going into space for over fifty years.

Now I am not saying we should create another cold war of some sort by putting space navies into orbit that are always at each other's throats and keeping us on the brink of another war from a global standpoint. But I am suggesting some types of weaponry and defensive systems in space make a whole lot of sense from my national and global perspective. We really need to rethink our policies when it comes to this.

Militarization of space is always used as some evil soundbite by politicians and activists and news anchors who don't really understand how our space program works. They don't truly understand how satellite communications, spy satellites, and missiles work. Perhaps educating the country a little better on everything we can and can't do in space and everything we would like to and need to do in space would go a long way in leading this elephant out of the living room and back into space where it belongs.

CHAPTER 13:
THE PLAN IN SUMMARY

In our first season of *Rocket City Rednecks* we worked really, really hard on projects and concepts that were fairly high tech that we managed to build out of low tech stuff. A lot of the tech stuff, like an Iron Man suit, we built out of junk parts. A season's worth of things piled up in the garage and our storage units. We most certainly would never throw any of that stuff away. First and foremost we had that RV we used for the Mars mission. Michael and Rog's first thoughts were "road trip."

So, the network decided to reward us with a trip somewhere. For us it was a no-brainer. We had to go to Talladega. I wanted to go and watch the cars push each other. Daddy wanted to go and hang out and talk to some of the people in the pit. Michael and Rog wanted to party. And Pete wanted to see the technology they used on the cars. When I told them we were going to get to go everybody was ecstatic. But it wasn't all fun and games. We still had to do a build. We had to make the ultimate tailgating weekend.

First things first; we had to make a plan of everything we'd need and want for the weekend. It was sort of like planning for a space mission. We had to have food and water. We would have to have shelter for the duration, and it got really cold at night down there. And we had to allow for unexpected contingencies, all while filming a TV episode.

We redesigned the vehicle and rebuilt it for the weekend. Pete invented a liquid nitrogen supercooler that would cool a hot beer to ice cold in four seconds. Five seconds and you had a beersicle. Rog built some nice grills and a windmill. Daddy and I built some grills, a sink, an automatically deployed cooking table, and a slide for fun.

We grabbed stuff we used throughout the season and found other uses for it. On the way down I started feeling a little weird. Then my head started hurting. Then it got really cold that first night. The first morning of filming at Talladega I was running a 103 degree Fahrenheit fever. That got me to thinking about frontier medicine for space travelers. As part of any future long-duration space missions we really need to think about how to handle illness and injury so far from home.

While the guys were out watching races and drinking cold beer I was eating cough drops and acetaminophen like popcorn. When we got to crawl around inside David Ragan's car and later when he came to see us, I was dang near delirious. The one thing I remember about David was how genuine he was. And how good at golf he must be. We had been putting on that upper deck golf course we made and none of us could hit worth a dang. David dropped a ball. Grabbed our homemade putter we called an "angle iron" because it was made from, well, angle iron. Then he putted once and put the ball right in the hole. I recall having visions of Daddy shooting that watermelon out of the sky with a rifle on the first shot. And, I hoped it was a good omen for David's chances in the race.

My feverish weekend continued right on through the race. I did manage to choke some medicine down, but that didn't seem to help. The adrenaline of the race did. David dang near won that race but with seven laps to go he ran over some debris that cut one of his spark plug lines. What if that happened to you between the Earth and Mars? You'd need a ship full of MacGyvers just to keep things going and not get you killed.

I never really got better until about two days after the race. Watching a rerun of that show I realized I could hear my voice was really strained. I was miserable that weekend but had the time of my life. That might be how astronauts feel in weightlessness or after long-duration missions. Most of them get God-awful motion sickness and have a terrible time while having the time of their lives. They are usually sick for days once they get home. Take the parallels of a weekend trip like this with high tech gadgets being implemented and then add an illness and on top of that add the danger of being in space miles and miles from anywhere.

Travelling in space is not easy. We are certainly not going to do it if we keep doing start up and stop and start up again approaches. We have to have a good long term plan like any good camping trip, vacation trip, or trip to Talladega. If you don't have it planned out for all the contingencies, an illness might pop up and ruin the trip for everyone, or you might break a spark plug wire and be stranded, or you might even get lost along the way. Without a good plan—a good, long-term, all encompassing plan—for where we are going in space, we have all gotten lost along the way. It is time we find our way back to being the once-great leader in spacefaring that we were and that we can be again.

Rednecks study an actual manned capsule trainer from the Mercury Program at the U.S. Space and Rocket Center in Huntsville, AL. Measurements of the Mercury capsule led to the design for the Double Barrel Rocket capsule.

Hopefully, some of the things we have done on *Rocket City Rednecks,* like building submarines, rocket powered boats, Iron Man suits, moonshine rockets, and whatever other crazy ideas we come up with to test and experiment with will stimulate other people's imagination and drive for doing. Doing. Doing. And more doing!

Part of our plan for doing this show was to help get people excited about doing science again. We wanted to show them that doing science is fun, exciting, and worthwhile. And it is important. It is important for the future of our nation and our species.

I plan on going back to Talladega. I'm planning on going to space. We should all be planning. We should all make good on those plans. Talladega is not so far away. Hopefully, our future in space isn't either.

★ ★ ★

So what should our national policy be on space exploration? Well, for a start . . .

Going to the Moon, or rather, going *back* to the Moon *is* a mission of national security, national morale, and economic security for the future of our country. If we did nothing to plan to go to the Moon, China, Russia, and India, likely in that order, will go to the Moon and take it for themselves. So what? What if they take the Moon for themselves?

Well, if we can figure out how to make it economically viable, the Moon should be filled with the same types of minerals and things that we find here on Earth. There should be gold and diamonds and gemstones and titanium and aluminum, uranium, you name it. Maybe that's not the immediate return on the national investment. The immediate return is: if we don't, those other nations that are developing the technologies to get to the Moon will be the smartest, most educated, and most technologically advanced cultures on the planet—not us. And if we decide not to do it ourselves, then we will lose our technological edge on the rest of the world. We will definitely not have another generation of high-tech scientists and engineers with marketable skills comparable to the ones China, India, and Russia will develop. And we will not have the spinoff technologies and economic windfall from that development effort.

And we definitely cannot expect any commercial entity to have the budget for an in-house independent research and development effort to build a launch vehicle to get to low Earth orbit, a trans-lunar injection vehicle to go from Earth to the lunar surface access capability, and then a return system to come home safely, including trans-Earth injection, reentry and landing. There is just no way that any known commercial entity can afford to do that mission. It's silly. It's not economically viable. Highly unlikely. It only makes for good political sound bites and that is all. Hey, I'm all for getting the government out of things they shouldn't be in on, but in a space program case, this is something that is of national importance, and the government must be the driving factor. In this case, the federal government is the only currently viable entity that can drive a real space program—the kind of space program that would make America greater than ever, again.

This literally is a state of national emergency. If we do not do a far-reaching space exploration plan centered mostly around manned spaceflight and to reach the Moon and Mars and beyond in the very near term, we will definitely lose our technological superior edge on the rest of the world. Our economy will suffer dramatically for a lack of space investment even worse than it already is.

Although the government will have to be the driving factor, the funding source, and the lead in going back to the Moon, we should definitely look for places where commercial competitions, reward challenges, and private investment will benefit the space exploration plan. An example of this might be use of the Spaceship One reentry system concept design. Others might be the Google lunar lander prize, the Centennial challenge to put small payloads into low Earth orbit in order to develop new, cheaper launch vehicle concepts. These types of challenges and contests need to have more than one simple prize. The competition will be too stiff for there to be one likely winner for there to be enough incentive to get a large mix of the nation's brain trust small engineering companies to go after these prizes. There should either be many awards for one challenge, or many, many challenges with single awards, or a mix of the two; otherwise, it would be unlikely to get broad-scope involvement in this competition.

If there is not enough budget to run programs, contests, and challenges in this manner, then those should be changed to extremely difficult concepts like faster-than-light travel and teleporters and maybe even anti-gravity, artificial gravity, tractor beams, deflector beams and shields, replicators, suspended animation chambers, artificial wormholes and stargates. Then there's unobtainium and unbelievium, which you then throw into an impossibility reactor to get impervium. You get the idea—things that are currently thought to be pseudo-impossible, but may not be. The challenge money for the lower-hanging fruit might be better spent in small business innovative research programs with multiple awards as opposed to one extremely difficult challenge that not many people will compete for. Michael Phelps couldn't push a barge off course but a thousand slower average swimmers working at it probably could.

At this point we can start laying down some general ideas about what a real national space strategy should be. At the center of the strategy, on top of it, would be the government entity whose mission is manned exploration of space and other space science missions. This of course would be NASA as it currently stands. NASA would be given the charter for the very large space exploration and development effort. Spinning off from NASA would be the commercial wing, whose job would be to purchase any available, commercially viable space access technologies for immediate near-term use. This should be limited only to American sources. Spending our money for foreign vehicles and other access to space technologies is not going to help invigorate *our* economy and grow *our* technology base *at all*. The commercial office would also have a large, strategic small-business innovative research effort. The scope of this should be expanded far beyond what is being done now at NASA. The SBIRs should be used for much more than new technology innovations. They should also be implemented to develop new components for near-term mission capabilities. This should stimulate small business across the country. The commercial wing could also run the contests and grand challenges if there are to be any.

There should also be another wing, perhaps a subset of the commercial, but perhaps just a wing in and of itself to foster the next generation of scientists and engineers and technologists at the university level and maybe even at the high school level. Programs like summer engineering apprentice programs, graduate assistantships, fellowships, scholarships, postdoctoral efforts, and independent research grants should be invigorated and improved and increased to make a way for the next-generation scientists to become involved in the process, and begin doing research, and learning *how* to do research that is viable for future space exploration.

Outside of NASA falls any and all private and commercial efforts not being funded by NASA and perhaps without even the intent to sell these capabilities to NASA. This would include tourism, space hotels, private access to space, and any other potential private endeavors. Regulations and taxes should be eased up for the private industry to incentivize them to do more space exploration technology development without particular government goals or end results in mind. There should be a wing of NASA whose *sole purpose* is to monitor growth in the private industry and learn from any innovations created there, leading back to potential purchases of the technologies for government use. Now all of these pieces exist today and/or have existed in NASA and the private sector. It is simply a matter of very little funding to keep them at a critical mass in order to be successful in growth.

Now, NASA already has NIAC—the NASA Institute for Advanced Concepts. It's the next-generation arm of NASA research. They're working on things like lunar colonies and missions to the outer planets. That's good. One other piece that we need: breakthrough physics propulsion for faster than light travel and warp drives. There should be a *next* next-generation research arm of NASA like DARPA. We need NIAC to become more invigorated and more of a super subagency for solving the hardest problems in space rather than an institute of tenured graybearded pipe smokers. Perhaps there should be a Space Projects Joint Technology Office venture between DoD and NASA to solve the hard space problems using both NASA and DoD dollars.

There is no Starfleet Academy. There is an astronaut program, but not much of one. With budget cuts and no launch vehicle available, the manned program within NASA has been gutted to near non-existence. Astronauts already in the Corps are moving on, looking for other jobs; applications have dropped off because . . . well, is it because young people believe there's no future in it without a launch vehicle? Or is it because they just aren't interested anymore? In the movies and books they've always made it look and sound so sexy and cool to be studying to be the next crew of space explorers for mankind who are preparing to "boldly go where no man has gone before." But it's not quite that glamorous—it's a heckuva lot of hard work.

Here is where we have missed the boat from a national standpoint. We have children's and even adult Space Camp, but this is nothing more than a space amusement park or vacation with a little bit of space education thrown in there. It's cool, and it's interesting, and you are gonna learn some stuff, but it isn't astronaut training. If you want to be an astronaut there is no known direct route to get there.

Okay, fine. What qualifies you to be an astronaut in the first place?

According to NASA's website,[24] "Any adult man or woman in excellent physical condition who meets the basic qualifications can be selected to enter astronaut training.

"For mission specialists and pilot astronauts, the minimum requirements include a bachelor's degree in engineering, science or mathematics from an accredited institution. Three years of related experience must follow the degree, and an advanced degree is desirable. Pilot astronauts must have at least 1,000 hours of experience in jet aircraft, and they need better vision than mission specialists.

"Becoming an astronaut is extremely competitive, with an average of more than 4,000 applicants for about 20 openings every two years."

Keep in mind that the requirements just stated are the absolute *minimum*. *No* astronaut gets into the corps with the absolute minimum, not with that kind of competitiveness. "The biggest challenge about being involved in the space program is the need to be able to be good at and know a lot about a lot of things," astronaut and Lieutenant Colonel Catherine G. "Cady" Coleman says.

Typical "other things" include master's degrees, doctorates, scuba diving, having a private pilot's license (if not being a military pilot), and the ability to pass the very stringent medical physical, with the following basic physical attributes:

- ❧ Distant and near visual acuity must be correctable to 20/20 in each eye,
- ❧ Blood pressure not to exceed 140/90 measured in a sitting position, and
- ❧ The candidate must have a standing height between 62 and 75 inches.

This latter is because of, among other things, space suits. Unlike

---

[24] http://www.nasa.gov/centers/kennedy/about/information/astronaut_faq.html#4

the early days of the space program, space suits are no longer custom made for the individual astronauts; they have been standardized, and the applicant must therefore be able to fit inside.

Let's take a look at the bio of someone who is already an astronaut. Michael E. Lopez-Alegria was born in Spain in 1958 but grew up in California. He graduated high school in 1976, got a BS in Systems Engineering in 1980, an MS in aeronautical engineering in 1988. He has another graduate degree (unspecified) from Harvard's Kennedy School of Government, specializing in national and international security. He speaks Spanish, French, and Russian. He is a Captain, U.S. Navy, Retired; a naval aviator, flight instructor, pilot and mission commander of EP-3E aircraft. He has been a test pilot and a program manager. He has more than 5,000 hours in the pilot seat of over thirty different aircraft. All *before* he applied to NASA's Astronaut Corps.

What about someone who doesn't have that military boost? How about Colin Michael Foale, who is a British citizen (now a dual British/American), born in 1957. He graduated from King's School, Canturbury (our equivalent of high school) in 1975. He got a BS in physics—with honors—from no less than Cambridge in 1978, and got his Ph.D. in astrophysics from the same university in 1982, then obtained a postdoc. In 1983 he joined NASA as a payload operations liaison at Johnson Space Center in Houston, TX and worked console for Space Shuttle missions STS-51G, 51-I, 61-B and 61-C.

He is fluent in Russian. His hobbies include wind surfing, private flying (fixed wing and helicopters), soaring, hiking, scuba diving, cross country skiing, and writing software—for fun.

He wasn't selected to become an astronaut until 1987.

If by some chance you already meet these sort of qualifications, application information can be found here: http://astronauts. nasa.gov/content/application.htm. Actual applications are submitted through http://www.usajobs.gov. Back in the days of paper sub-missions, an astronaut application amounted to a small book, and could be submitted at any time and held on reserve until needed. Now calls are put out and applied for online. The next class, as of this writing, is anticipated to be in 2013. (Keep an eye here: http://astronauts.nasa.gov/ if you're interested.)

But what if you aren't qualified yet but hope to be? How on earth do you manage to get competitive qualifications otherwise? Well, these guys and gals are overachievers by nature; they pack a lot of activity into a day. But if there were an academy, like I said, that would eliminate the competition coming out, because if you passed the space fleet academy, like any military academy, you would have passed the requirements. The competition then becomes getting in and getting through—but that's as it should be.

The American Military University is offering accredited online Bachelor of Science and Master of Science degrees in Space Studies[25] and it is open to civilians as well as the military. That's a start, but it's far from being enough for our purposes. Online studies can only provide so much. Hands-on studies, lab work, simulators, and the like are needed too. That means we *need* a specific facility.

What we need is a full up academy. We need a NASA/U.S.A.F./Army/Navy/Marines/Civil Space Academy. A centralized university type environment (I'd prefer it be in the Rocket City but I'm biased) where young people can go and get degrees in Space-Oriented Science, Engineering, and even Management. The curriculum should contain every aspect from astrophysics to rocketry to pilot training to mission control to medical training and anything else in between. It is at the United States of America Space Fleet Academy where anybody who can get accepted (through standard collegiate requirements and pilot physical status) can go to this university and get his or her space education. The degrees would all be tailored to land space jobs for graduated students. Pilot trainees could spring-board into military, civil, or commercial aviation for further training and maintaining currency in pilot status. The other students (non-pilots) would acquire Bachelor of Science, Master of Science, and Doctor of Philosophy degrees in their particular fields.

The output of the academy would be a generation of young people fully trained and full of enthusiasm to push forward into space and to bring on a new era of humanity's involvement in space. The more students we educate on the benefits of space exploration and

---

[25] http://www.amu.apus.edu/lp2/space-studies/bachelors.htm

utilization, the more mainstream it will become in our culture, society, and business.

Private companies and university activities really cannot generate the pull themselves to create real space exploration programs, with real national, and even global, attention. The USA Space Fleet Academy would. It would be easy for such an organization to be set up with modest amounts of the national space exploration budget. If the budget matched what it should, compared to that of the Apollo era, it would be *very* easy. Five percent of the NASA budget could easily set up such an academy, with some to spare. Since the other services and industry would be involved, they could also become partners in the expense. Five percent of the space exploration budget currently is somewhere between $500 million and $1 billion a year. This could set up flight training capabilities, space mission simulators, laboratories, classrooms, world-class instructors, and even the more mundane logistical pieces of such an academy such as buying a campus, setting up buildings, and marketing to students.

There are pieces of the space program, like such an academy, that we have simply overlooked. If we train a Space Fleet, then soon enough that fleet will want to go into space. And they will get frustrated with the politicians if it doesn't happen. We will also train a large group of individuals on how important it is for humanity to stretch out into the heavens. That group of people will then communicate with their families and friends and spread the word. The concept of humanity in space would become as viral as being "Army Strong," or pulling for the New York Giants to win a Super Bowl.

We've talked about the generalities of America in space. Talked about how NASA has done things in the past. We've talked about how the private industry has done things in the past. And we have given the general idea of the way it should be approached for the future. But what about the details? How should we really get back to space and have that future we were promised back in the '60s and '70s?

One of the first things I would have done is left Mir in space all those years ago. I would have had private industry enthusiasts see if they can do anything with it. Or leave it there and use rights of salvage for whoever could get there and use it. Why go through the hassle to deorbit the thing and burn it up later due to political pressure?

It really all came down to politics. One lesson we will have to learn is not to throw stuff away that we already have in space. When we convinced the Russians to deorbit Mir, I thought that was the biggest mistake we could ever make. Not because Mir was salvageable as far as a space station was concerned; it was falling apart. But Mir was already in orbit. At $10,000-$20,000 per pound to low Earth orbit—and Mir weighed many, many tons—it didn't make sense to me to bring those materials back down to Earth and then pay all over again to take stuff back up. A little bit of thinking could lead us along a path to find a use for the materials that made up the Russian Mir space station. The solar panels might not work very well, but they did work, and we could've used them. If we needed an extra bulkhead on the back porch of the International Space Station, or a panel to patch a micrometeorite hole, or whatever, we could have gotten it from the Russian space station. But we can't do it anymore since we deorbited it.

Granted, I probably wouldn't want to use anything with potential flammability, or electrical shock hazards, or pieces that might fail, as a critical structural component on the International Space Station. But certainly we could've expanded the back porch for the space station for experiments to go outside and not in a protected environment where the astronauts stay. Or if we didn't want it for the International Space Station we might have sold it to some private commercial organizations. They might have found a use for it. Or we could simply have left it in place for salvage rights, and whoever could get to it and wanted or needed to get it could if they were able. We might turn the thing into a low Earth orbit refueling station. Could've just been a platform to which we'd attach supplies to come by and get later. Now the orbit wasn't exactly right; we could have fixed that. It could've been a platform for college-level experiments. Maybe targets for the various defense departments of the world to use for something. But that might be the weaponization of space, and God forbid that.

So the bottom line is, if there is stuff there that is not causing a hazard to someone in particular, leave it there and reuse it. Just food for thought: how hard would it have been to install a trans-lunar injection thruster on the Russian space station and park it around the Moon? Or even just crash it into the Moon so that we have a scrap pile there to use later.

★ ★ ★

First process is to get the astronauts a way back into space immediately. The fact that we have to rent a ride from the Russians to get to our own space station is ridiculous. The shuttle should be pulled out of mothballs immediately and be ready to go within this year to fill in the gap until we have a shuttle replacement. Any other solution is just plain stupid. If you've got an old clunker station wagon, and you have to get your kids to school and the bus doesn't come all the way to your house, you don't go and put your clunker in the junkyard until you can buy a new car. Whoever's idea this was, well, it wasn't a good one.

Unfortunately, the shuttles were gutted before being sent to museums. They didn't get sent whole. The parts are scattered all over the damn place. But there still has to be enough to put one of them back together quicker than building a whole new rocket program. And if not, this is an extreme disservice the current administration has done to America. We should look for an immediate to very near-term solution that doesn't include Russia or China.

And instantly we should be man-rating another vehicle so we *can* retire the shuttles. The Ares-1 vehicle from the Constellation program was on the way. It was an extremely low-budget program, so if the budget was made appropriate, it could be accelerated and probably replace the shuttle within about two years. The Obama administration decided, even after we had done a successful flight test of the vehicle and spent billions to develop it, that it needed killing because President G.W. Bush started the thing.

Another near-term option would be to take the space launch systems concept and slim it down. Take a Delta-4, and I mean a Delta-4 based on American rocket engines and not an Atlas-5, which uses a licensed Russian rocket engine, and go through the process to man-rate it and add a smaller spacecraft to carry only two or three astronauts to the space station. This should be a good backup program anyway and we ought to have this capability. This is easier said than done, though, because the Delta-4—and Atlas-5 for that matter— were not designed to be man-rated to carry astronauts.

We should also consider beefing up our budget to Elon Musk's organization and have the Falcon-9 testing accelerated in order to man-rate the launch vehicle and his capsule designed for holding two to three astronauts and carrying them to the space station. Then we would have the potential for the Delta-4, Falcon-9, and then at that point maybe the Atlas-5 as well, carrying astronauts back and forth to the space station and low Earth orbit. The Ares-1 would probably come online about the same time we could get the expendable launch vehicles man-rated. Then we would have four new solutions and could retire the space shuttles from carrying astronauts.

Other than the manned space missions, any good American space plan should contain details of unmanned and terrestrial-based exploration of space. Although the lion's share of the budget is likely to be taken up by the manned space program component, a significant amount of funding should be available for terrestrial-based exploration of space, and far future unmanned probes. The really far off outer planets, and even some of the inner planets, in our star system are either too far away or have extremely harsh environments, such that our technology does not allow us to send manned explorers there. We can send unmanned vehicles to some of these locations, and much more near-term missions, while we wait for the technology to advance to the point where we could send humans there. A good example of this would be the Pluto-Kuiper Belt object New Horizons mission. The mission duration is on the order of ten to twenty years because Pluto is so far away. There isn't propulsion technology or suspended animation or any other technology known to man that would allow for humans to withstand such a long space flight at this time. But scientifically learning about what Pluto is made of, as well as what the other objects in the Kuiper belt are made of, is extremely important scientifically. We need to understand as much as we can about all the planets out there in order to create a good model of our solar system's origin.

We could also use the unmanned space probes for forward reconnaissance before we do manned missions to the Moon or to Mars. For the Moon we need very detailed high-resolution 3-D maps to determine the best place for our new lunar colony construction site. We need a lunar GPS satellite constellation. We also need a lunar global communications system. The unmanned spacecraft will either serve an application or be sent to solve scientific questions. The unmanned probes do not necessarily need to go to a planetary destination.

Some great examples are the Hubble telescope, the Chandra X-ray Observatory, the Midcourse Space Experiment, the Kepler planet finder spacecraft, and Gravity Probe B. All of these spacecraft and many more have helped us find detailed information about the universe, our solar system, our planet, and ourselves within it that we could not have discovered without the space missions. The science missions, although not as big in demand as manned space missions, still put a lot of people to work *and* solve a lot of scientific questions that couldn't have been solved without them. These missions put a lot of scientists and engineers and analysts to work designing and testing, constructing the spacecraft, and implementing them, in performing the science experiments, and analyzing data for years to come after the missions. We need to create a list of science missions that need to be done in the near term to solve problems with near-term applications, as well as a list including large gain science missions. One example of the large gain science missions is the Kepler planet finder. We now know that planets are out there around almost every star and we're even beginning to realize there might be many Earthlike planets—some as close as 22 light-years away. Another example of a high science return mission is the mission to develop major aspects of our understanding of gravitation like the Gravity Probe B. Someday understanding these things may enable us to create advanced propulsion concepts like we see and read about in science fiction movies and books.

The terrestrial-based space program consists of land-based radio telescopes, optical telescopes, and even radars and laser range systems. We need to beef up our survey of the night sky for near Earth objects, asteroids and stuff like that. We need to offer incentives for amateurs to purchase better telescopes and planet-finding and asteroid-finding equipment. Since this is a service to our country and mankind in general, it ought to come with a tax break, too. Our national observatories and observatories in academia should have their budgets, plus an addition, to modernize and to train our next generation of astronomers, astrophysicists, and cosmologists. And now that we know where planets are, as we find them we should create and renew the search for extraterrestrial intelligence. The government has not funded SETI for decades. But the search for extraterrestrial intelligence used to be more of just a shot in the dark. Now, as we find potential

Earthlike planets, we should focus our efforts in SETI on those planets. This should be a government-funded project, not just an enthusiast project. There is a bundle of scientific information to be had—if we know there is a planet out there that is patently able to harbor life, then we should be studying that planet to determine if life *is* there. Think about it, who do you want making first contact with aliens? Shouldn't it be from an orchestrated group of folks with a set of contact rules approved and put in place? If amateurs do it, there will be no such regulations to govern them.

With the advent of modern adaptive optics technology and high-speed computing, we're learning how to remove atmospheric turbulence issues between our Earth-based telescopes and space. Modern technology would allow for us to build very large telescopes tens of meters in diameter on Earth with adaptive optics systems that would allow us to remove the effects of the atmosphere. These types of telescopes might help enable us to study and image extraterrestrial planets—those in our own solar system—and extra-solar planets.

We've already started this sort of program. Has it been worth it?

The Hubble Space Telescope (HST) had been planned since 1970, but in 1974 its budget was zeroed out. After a national lobby in the space and astronomical community revived it, it was re-funded, but at a smaller cost. In addition, the European Space Agency (ESA) threw in funding in exchange for at least fifteen percent of the observing time. The original estimate for the program was some $400 million, and this was approved by Congress in 1978, with an allocation of $36 million for that year. MSFC and GSFC were given joint oversight, and the work was contracted to Perkins-Elmer and Lockheed. Perkins-Elmer had the all-important mirror as part of its construction work. Launch was set for 1983.

Almost immediately the oversight organizations began to have concerns over the contractor management. The primary mirror, in particular, slipped behind schedule and over budget. The launch was slipped to 1984. NASA still noted contractor issues, and the launch date slipped to 1985, then 1986. At this point the cost had ballooned to $1.175 billion. Lockheed, responsible for the housing in which the telescope would be placed for its ride into orbit, had problems as well. MSFC noted that the housing was thirty percent over budget, and three months behind schedule. In addition, they felt the Lockheed contractors lacked initiative, and instead of proceeding, tended to wait for specific NASA direction on all points.

An additional delay occurred as a result of the *Challenger* disaster and the grounding of the fleet. Eventually, after Return To Flight, HST was launched on 24 April 1990 aboard shuttle *Discovery*. It was placed into low Earth orbit. Final cost to launch: $2.5 billion.

HST was intended to view the visible spectrum as well as ultraviolet (UV) and the near infrared (IR). It had a forty-eight-square foot (4.5 m^2) mirror, and a focal length of 189 ft (57.6 m). Almost immediately it was discovered there was a problem with the primary mirror—spherical aberration, a condition in which the mirror is not shaped in the correct parabolic curve. However, even first light proved sharper than most Earth-based telescopes. Initially, software was used to correct for this not-uncommon, but still unwelcome, condition. Eventually corrective optics were taken up and installed in a subsequent shuttle flight. Since its launch, periodic shuttle visits have maintained and occasionally updated the HST, and GSFC and the Space Telescope Science Institute (STScI) have provided ground support. This has resulted in a cumulative cost of somewhere around $4.5–6 billion for the U.S., and an additional €539 million from ESA.

Was all of that worth the effort?

The discoveries that HST has made to date include determining the most accurate distances to Cepheid variables ever made; constraining the Hubble constant (an important value in determining the size and age of the Universe) from previous error bars of ±50% to ±10%; detecting the fact that the Universe's expansion is accelerating, not decreasing (and thus leading to the theory of dark energy); observing the impact of Comet Shoemaker-Levy 9's pieces with Jupiter; and ascertaining the high probability of a supermassive black hole in the Milky Way's galactic core. The list of its discoveries would probably fill this book. It has been functioning for over 21 years and is expected to keep going until at least 2014. Its successor, the James Webb Space Telescope, is expected to launch in 2018. All in all, not too shabby a result. Might've been done better and cheaper if the "commercial" aspect had been taken out of the equation, actually. And my biggest concern right now is why are we waiting so long on the James Webb Space Telescope and how the heck are we going to launch it into space? The current plan for launch involves foreign launch vehicles. Go figure.

The Chandra X-Ray Telescope was originally dubbed AXAF, for Advanced X-ray Astrophysical Facility. It was first proposed in 1976, and the work was to be done by MSFC and the Harvard-Smithsonian Astrophysical Observatory (SAO). It was slow getting off the mark, however, for the simple reason that the people involved were more than aware of the cost overruns occurring concurrently on Hubble, and they didn't want to present a proposal to the upper echelons that might get perceived as "Hubble II."

Eventually it got into the race. But in 1992 the project was forced to redesign the telescope due to cost-saving measures. This resulted in a reduction in capability of the telescope. Four out of a dozen mirrors were eliminated, and two out of six experiments were cut.

It was planned for a high, extremely elliptical orbit that took it a significant portion of the way to the Moon, which meant that we had nothing, not even a shuttle, that could maintain it. On the other hand, it also put it above most of the Earth's radiation belts, which would interfere with its operation by producing excess noise in the data. It was launched aboard Columbia in 1999, with a planned lifetime of five years. In 2001 this was formally extended to ten years. It is currently at a mission elapsed time of over twelve years and counting, with the capability of operating for at least fifteen years. Its successor, the International X-Ray Observatory, was canceled in 2008. Probably wouldn't have been able to get the right power cord adapters for the thing anyway.

Chandra, on the other hand, has an aperture of 0.43 square feet (0.04 m^2) at an energy of 1 kilo-electron volt (electron volts are units of energy used in particle physics and photonic calculations—it is extremely tiny and in this case represents the energy of a single X-ray of a particular frequency). It has a focal length of thirty-three feet (10 m).

Overall development and launch costs were roughly $2 billion. Ground ops and data analysis in its first ten years of operation ran roughly $1 billion, with the second five years costing approximately one third what the first 5 cost, indicating a streamlining of mission operations.

Discoveries made by Chandra include the fact that the Crab Nebula pulsar has a ring around it whose dynamics are still being debated; the first detection of X-rays from Sagittarius A, the supermassive black hole at the Milky Way's core (that HST verified); the fact that Perseus A is cannibalizing a dwarf galaxy; black holes that are the "missing link"—intermediate in mass between stellar black holes and the supermassive black holes that form galactic cores; and possible quark stars. We've learned a lot of cosmology and astrophysics with Chandra and we need to continue and step up these types of missions.

The Midcourse Space Experiment (MSE) is a Ballistic Missile Defense Organization satellite, operating in the far infrared, visible, and ultraviolet spectra. It was intended to help track ICBMs and to determine what objects that any ICBM trackers should avoid as false targets. It was launched on a Delta-2 from Vandenberg Air Force Base in 1996. It was placed in what is known as a sun-synchronous orbit around Earth, at an altitude of 558 miles (898 km). This is a near-polar orbit intended to maintain the same relative Sun/Earth angles throughout.

Due to its status as a DoD sensor, little is known about the budget.

It provided an excellent mapping of the galactic plane in infrared wavelengths. In fact, there is a common map of the night sky at infrared wavelengths that was generated by MSX. From this map we learned a lot about our universe like where it is hot and where it is not. This great data came from a weapon program. Again, there is nothing to the myth that space is not to be weaponized. In fact, it is a bit stupid for us not to do so unless we want to become extinct like dinosaurs some day.

The Kepler planet-finding mission is one of NASA's Discovery missions. Such missions are cost-capped proposals, with a principal investigator in charge who is expected to maintain the budget with no overruns. Such missions are open to industry, small businesses, government laboratories, and universities. In this case the participating groups are NASA's Jet Propulsion Laboratory (JPL) and Ames Research Center (ARC), and Ball Aerospace. The year 2006 saw a launch delay due to budget and fiscal concerns, as well as a redesign of the high gain antenna (gimbal to fixed mount) to save money. This, however, resulted in the operational loss of one observing day per month. The total cost of Kepler has been less than $425 million.

It was launched in 2009 with an expected mission life of at least three and a half years. It is currently about two thirds through that expected lifetime and still going strong. In 2010, two out of a total forty-two charge-coupled devices (CCDs, electronic light sensors used in digital cameras and telescopes) failed. This was not considered significant, and the mission continued. Mission operations are handled out of the Laboratory for Atmospheric and Space Physics (LASP) in Boulder, CO.

Designed to survey a limited area of sky in detail, Kepler studies the constellations of Cygnus, Lyra, and Draco, containing over 145,000 stars, looking for the faint signs of variability that indicate a planet might be passing in front of one of them. To date, it has not only identified many previously unrecognized intrinsic variable stars, but has found 2,326 planetary candidates, including 207 Earth-sized; 680 "Super-Earth" sized (1–2x Earth mass); 1,181 Neptune sized; 203 Jupiter sized; and 55 larger than Jupiter.

Statistical results of this survey already indicate that some 1.4–2.7 percent of all sunlike stars are expected to have at least one Earth-like planet in the so-called Goldilocks zone enabling habitation by Earthlike life forms.

Gravity Probe B was an effort to essentially map the Earth's gravity well and verify Einstein's space-time gravity well geodetic effects (how the Earth's mass affected the shape of space-time, and how it might be dragged around Earth with the planet's motion) as predicted by General Relativity. Previous efforts had measured these effects with an error of ten to twenty percent, but GPB intended to reduce that to one percent.

The concept was first developed in the late 1950s. It was proposed to NASA in 1961. NASA approved a grant in 1964 and supported the research until the grant expired in 1977, at which time engineering designs had been made. In 1986 they found they would not be allowed to launch the probe aboard the shuttle. This necessitated a redesign in order that the probe could be launched on a Delta-2, and in 1995 they were denied the opportunity to fly a prototype aboard the Shuttle as well.

Its total cost was $760 million. It was launched aboard a Delta-2 from Vandenberg AFB in April 2004 and decommissioned in December of 2010. The Delta rocket placed it into a 400 mile (642 km) polar orbit, where it remains.

In early 2007, ground controllers realized there was a problem: the superconductor gyroscopes' spin axes were being affected by torques which were varying with time. This was, later that same year, determined to be due to electrostatic forces within the gyroscopes' spheres' coating, which created an unexpected potential drop across each sphere, creating a classic dipole torque. It also induced currents in the housing, which resulted in the torque changing with time.

In addition, there were forced breaks in data collection due to solar flares.

As a result of these problems, NASA withdrew funding in 2008, but the team secured alternate funding from the King Abdulaziz City of Science and Technology (KACST) in Saudi Arabia, which enabled it to complete its full mission. Again, we had to go outside of America for what should have a major part of America's space program.

In May of 2011, the Gravity Probe B team announced that it had succeeded in verifying Einstein's two predictions to the desired level of accuracy. However, due to the gyroscopic problems, these results are currently being debated in the astrophysical community.

It is from Gravity Probe B that we are beginning to get a better understanding of some of the more esoteric aspects of how gravity works. Newton was only close when the apple fell on his head. Einstein suggested there was a deeper explanation and understanding to be had but couldn't prove it. Gravity Probe B helped prove that Einstein was right. And now, thanks to that mission, we are beginning to peek behind the curtain at what gravity might really be, how it really works and why we might be getting closer to understanding. Until we fully understand gravity things like warp drives and artificial gravity will continue to be somewhat elusive, to say the least. We *must* have more experiments like Gravity Probe B.

We've learned a great deal from our past unmanned space programs that have actually changed how we think about the universe.

It's long been suspected that one or more large water oceans once existed on Mars, particularly in a large depression centered on the Martian North Pole. Actually, all of the surrounding, elevated features show signs of erosion and fluid flow. This depression is tentatively dubbed Oceanus Borealis. Very recent discoveries by sensors on ESA's Mars Express orbiter indicate that not only is the suspicion likely true, but that particularly around the southern pole, there may be large amounts of water ice. In the northern oceanic basin, it is less clear-cut, and may represent desiccated water sediments, or water sediments laced with "massive ice."[26] Other observations indicate the presence of thick brine seeps, where the water is mixed virtually one to one with salt.

If it's possible that life that is not water-based exists, it might likely live on Titan, one of the Moons of Saturn. We discovered with the Huygens probe that Titan is covered with lakes of liquid methane. On Titan it rains liquid methane, but the surface is mostly sub-zero frozen water ice. Since there isn't a sufficient source of methane on the surface to account for its abundance in the atmosphere, it has been proposed that it is replenished geologically. This leads to a concept that is called "cryotectonics." The volcanoes don't spew molten rock. Instead, they spew water, methane, and ammonia. This little planet—yes, planet, it's larger than Mercury—is covered with methane clouds, and has an atmosphere far thicker than Earth's, but like Earth's, composed principally of nitrogen. The high concentration of organics in the atmosphere (formed by reaction of sunlight with the methane) made scientists think that there would be precipitated organic sediments on the surface, but instead spectral analysis found water ice, leading scientists to wonder: where are the organics?

We don't even know at this time if the surface is smooth or rough. If it's rough, the organics (hence the possibility of life) would tend to collect in the "valleys." This latter type of surface is considered likely, as the rate of erosion is estimated at only one four-hundredth that of Earth, and mountain building (tectonics) is occurring at one fiftieth Earth's rate. So, hills and mountains should be building

---

[26] http://www.skyandtelescope.com/community/skyblog/newsblog/New-Evidence-for-Ancient-Martian-Ocean-139085114.html

much faster than they are being eroded away, and will be very exotic, featuring mountains and craters and possibly very little in the way of lowlands.

On the other hand, in 2001 the Arecibo radio observatory in Puerto Rico sent powerful pulses of radio energy toward Titan, using the radio telescope as a giant radar, and got what are called specular, or mirrorlike, reflections off large areas of the surface. This had to come from regions so flat they could only be the surfaces of broad areas of liquid. The region observed was a swath parallel with Titan's equator, centered on latitude 26° south, and extending for an arc of about 22° in longitude. In this area, they got specular reflections about seventy-five percent of the time, indicating a large network of lakes and large bodies of liquid. In addition, the Cassini spacecraft discovered a large flat area at Titan's southern pole that is likely a methane ocean, judging by spectral data. The Huygens probe that it dropped photographed river systems as it descended, and landed in what was, essentially, muddy ground.[27]

Given the presence of tectonic activity, water ice, and organics, Titan may be the most likely source of life outside the Earth in our own solar system. But then, there's still Mars and Europa.

Europa has a deep, maybe as much as sixty miles deep, ocean on it. There are heavy gravitational tidal forces acting on it due to it being one of Jupiter's moons. The large planet tugs at the icy moon constantly. The ocean floor is likely to have volcanic vents and lava tubes in large quantities. Here on Earth at those locations there is chemosynthetic life. We need to go there with unmanned probes to have a look.

In the past most of our science missions have been designed as one of a kind. Making spacecraft that were custom jobs for each mission in this case made them fairly expensive. Clearly most science missions will have specific requirements that need custom equipment. However, it is likely we could develop a modular set of components for future space missions or unmanned probes. An inventory of

---

[27] http://www.skyandtelescope.com/news/3307256.html
http://www.skyandtelescope.com/community/skyblog/newsblog/26152854.html

modular components that snap together to perform whichever mission is required might make the cost of the science missions much cheaper in the long run. Imagine a box of spacecraft Legos and all we have to do is define the requirements of our space mission and then pull out the right Lego blocks and snap them together and we're ready to go. Of course there could always be a platform with a universal bus for accepting custom sensors and equipment.

Another aspect of unmanned efforts might be to place a radio observatory on the far side of the Moon. We can't observe a lot of the radio universe right now simply because Earth is so noisy when it comes to radio waves. On the far side of the Moon we would have the entire Moon blocking out and attenuating that background radio noise coming from our own television and radio broadcasts. This might enable access to an entirely new spectral band and usher in a new era of radio astronomy. This could be as simple as a lander radio telescope, or one that only functions as a radio telescope once it's on the far side of the Moon in its orbit and then sends data back to Earth when it's on the Earth-facing side, or it could also be an extension of the lunar colony. There could be an outpost on the far side of the Moon that is an extension from command. Naturally occurring craters make the perfect sites for radio telescopes patterned after the Arecibo telescope in Puerto Rico, and come in sizes to suit need, desire, and construction ability, up to and including the sixty-two-mile (100 km)-diameter Daedalus crater.

Problems with a lunar surface based radio telescope arise with the fine, abrasive lunar dust, however. The components of the telescope—especially the moving ones—will have to be protected from this dust. The lack of atmosphere is a help, because the only source of such dust will be the construction workers themselves, as well as, possibly, the occasional impactor.

Another alternative is the L2 point. This Lagrange (or libration) point is one of five similar points that are regions of relative gravitational stability in a two-body system. (The trans-lunar injection trajectory typically intersects the return trajectory at a Lagrange point, where the Earth's gravitational force equals the Moon's on the spacecraft.) The L2 point for the Earth/Moon system is about 39,000 miles (62,800 km) past the Moon's far side, straight out on a line between the cores of the two bodies. Placing a radio telescope at this location means that it can take advantage of the lack of dust contamination and the shielding from artificial signals provided by the bulk of the Moon, and still be constructed as big as our capabilities allow. In addition, because of the nature of the L2 point, the telescope system can be placed in a Lissajous orbit (a quasi-periodic, non-elliptical orbit) that is sustainable with only a minimum amount of station keeping.

Whether surface, orbital, or space-based, these telescopes will have to be protected from solar flares and coronal mass ejection (CME) events that could not only damage electronics but also induce charges in the very structure of the reflector.

The exciting part about space-based or lunar-based radio astronomy is that it opens up entire portions of the radio spectrum that Earth-based radio telescopes can't see, either due to their selection as bands for human transmission, or because of atmospheric absorption. This includes any radio wavelengths longer than about sixty-five feet (20 m), because the Earth's ionosphere either absorbs or reflects these radio waves almost completely. This section of the spectrum may contain important data about the cosmic background radiation that has the potential to change our very outlook on the universe itself.

We can also look at the microwave region (which is totally filled with manmade noise on Earth because of our own emissions) which includes the very special twenty-one-centimeter band of hydrogen emission. Since this is a strong and obvious line in that region of the spectrum, it is often argued that any extra-terrestrial intelligence would use it, or the spectral region around it, as a carrier for a communications signal. But it is just as likely that any ET signals are simply signals that the alien cultures use for communications, which may or may not be in bands that we have filled with manmade noise. This is a reason for putting observatories on the far side of the Moon. The detection of an alien signal would be a monumental, world-wide event even greater than the Apollo 11 lunar landing.

Or perhaps we won't find the aliens in the near future. There is still plenty of science, astronomy, and engineering to discover in space. We must go back to the Moon as soon as we can to enrich our nation with the technical, economic, and emotional boost that we so desperately need right now.

This new space plan is to do it all in logical and calculated steps, but with enthusiasm and vigor and the drive of a moral imperative. Our moral imperative is to make and keep America strong, safe, and bountiful and to protect humanity's future.

The approach is to be smart about our plan and not throw away a legacy we've been building for decades. It is humanity's destiny to be amongst the stars, and Americans should be leading every step of the way.

I'm from the Rocket City where a handful of former Nazi German scientists and a town full of sharecropping redneck farmers showed the world that man could and should be in space, on the Moon. My father milled out parts that flew on those great rocket ships that shook the Earth and gave us the Moon. My generation has lost sight of that, but not for much longer. If we all dig deeper and stay with it we will get ourselves back on track and back in space. Where is Captain Kirk when we need him? Well, we've got an entire town full of second and third generation redneck farmers turned rocket scientists and we are sitting on ready, and waiting to blast off. It is time, America, for a town full of Captain Kirks to stand up. It is time for a nation full of Captain Kirks to stand up. It is time for us to take back our space heritage and legacy.

I know that those of us here in the Rocket City are ready to go back to space with all our hearts and souls to pour into the endeavor. I know that we all have our own ideas of how we should be doing that. Well, it is time for us to fish or cut bait and decide what that plan is. Whatever the plan, we must stick to it and not let the next election kill it for political spite. We must find a means to stop politics from killing America in space. That is the key to America's success in space and pretty much everything else. The main part of the plan is to pick a plan and STAY WITH IT!

This is my plan for America in space. This is a New American Space Plan.

I am Dr. Travis S. Taylor, Ringleader of the Rocket City Rednecks, and I approve this message!

Appendix: Timeline of Space-Related Events		
Year	**Country**	**Event**
1926	USA	Robert Goddard launches his first rocket; liquid fueled
1933	USSR	GIRD launches GIRD-09; liquid fueled
1933	USSR	GIRD launches GIRD-X; hybrid fueled
1948	USA	Bell X-1 breaks Mach 1
1951	USSR	In individual flights, 2 dogs successfully put in suborbital exoatmospheric trajectories
1957	USSR	First ICBM, R-7 Semyorka
1957	USSR	Sputnik 1
1957	USSR	Sputnik 2; first animal (dog—Laika) placed into orbit. Died within hours from overheating due to component failure.
1958	USA	Explorer 1—first U.S. satellite; discovers Van Allen belts
1958	USA	Vanguard 1
1959	USSR	Luna 1—lunar flyby
1959	USA	Pioneer 4—lunar flyby
1959	USSR	USSR
1959	USA	X-15 flights begin

Year	Country	Event
1959	USSR	Luna 3 provides first lunar farside imagery
1959	USA	Little Joe 2—first American animal (rhesus monkey) flown; suborbital, nonexo trajectory
1960	USSR	First animals (dogs) successfully returned from orbit
1960	USSR	Nedelin catastrophe—second stage ignites first stage on pad, tremendous explosion, release of toxic fumes. Official death toll 78; presumed death toll 120. One survivor, who had gone into bunker for a cigarette.
1961	USA	Mercury 2—carries chimpanzee into suborbital, exoatmospheric trajectory
1961	USSR	Venera 1
1961	USA	Ranger 1—lunar test flight
1961	CSA	First launch of Black Brant rocket (suborbital)
1961	USSR	First cosmonaut dies in space capsule fire during ground tests in low pressure, 100% oxygen environment
1961	USA	Mercury-Atlas 5—carried chimpanzee into orbital flight
1961	USSR	Vostok 1—Yuri Gagarin first human in space; orbital flight
1961	USA	Mercury 3—Alan Shepard first American in space; suborbital flight

Year	Country	Event
1961	USA	Liberty Bell 7—"Gus" Grissom nearly drowns when hatch blows prematurely on splashdown; spacecraft sinks in ocean. Suborbital flight.
1962	USA	Mercury 6—John Glenn first American to orbit Earth
1962	USSR	Vostok 2—24+ hours in space
1962	USA	Mariner 2—first successful planetary encounter—first successful flyby of Venus
1962	CSA	First Canadian satellite, Alouette-1, launched
1962	USSR	Vostok 3–4 dual flight
1962	USSR	Mars 1 probe
1963	USSR	Vostok 6—Valentina Tereshkova, first woman in space
1963	USA	Faith 7—first American to spend 1 day+ in orbit. Last solo space flight.
1964	USSR	Voshkod 1—first 3-man crew
1964	ESRO	European Space Research Organization founded
1964	USA	Mariner 3—failed Mars flyby
1964	USA	Mariner 4—first Mars flyby
1965	USA	Gemini 3—first manned Gemini flight

Year	Country	Event
1965	USSR	Voshkod 2—first spacewalk (not entirely successful)
1965	USA	Gemini 4—first American spacewalk (successful)
1965	USA	Gemini 5—first manned craft to spend one week in space
1965	USSR	Venera 3—first impactor on another planet
1965	USSR	N-1 lunar launch vehicle begins development
1965	USA	Gemini 6a–7—first American rendezvous; Gemini 7 sets space duration record of 14 days
1966	USSR	"Chief Designer" Sergei Korolev dies
1966	USA	Gemini 8—first manned docking to unmanned spacecraft; thruster malfunction created near-unrecoverable tumbling upon undocking. Astronaut Armstrong corrected; first emergency landing of a manned U.S. mission.
1966	USSR	Luna 9—first soft landing on another celestial body
1966	USSR	Luna 10—first established lunar orbit
1966	USA	Gemini 12—"Buzz" Aldrin sets EVA record 5 hrs 30 min
1967	USA	Apollo 1 fire—astronauts Grissom, White, and Chaffee lost

Year	Country	Event
1967	USSR	Soyuz 1—first loss of cosmonaut during mission, when everything that could go wrong did
1967	USSR	Cosmos 186–188—first unmanned rendezvous and docking
1967	USA	Apollo 4—first full-up test flight of entire Saturn V stack; nearly shakes apart Walter Cronkite's broadcast booth 4 mi (6 km) away
1968	USA	Apollo 5—first orbital test of LM; unmanned
1968	USA	Apollo 6—unmanned; intended to send LM mockup into trans-lunar injection. Vibrations caused premature shutdown of 2 2nd-stage engines; third stage did not fire. Trans-lunar injection unsuccessful.
1968	USA	Apollo 7—first flight of redesigned command module since Apollo 1 fire; Earth orbital only
1968	USA	Apollo 8—orbits the Moon without landing during Christmas holidays
1969	USA	Apollo 11—first manned lunar landing
1969	USSR	Soyuz 4–5—first double manned craft docking and crew exchange
1969	NASDA	Japanese space agency NASDA established
1970	USSR	Luna 16—first unmanned soil sample return

Year	Country	Event
1970	USSR	Venera 7—first successful Venusian landing
1970	NASDA	First Japanese satellite, Osumi, launched
1970	USA	Apollo 13—major malfunction (O2 tank rupture in Service Module) aborts lunar landing; considerable ingenuity used to prevent loss of crew
1970	USSR	Lunakhod 1—first unmanned rover
1971	USSR	Salyut space station program begins (multiple Salyuts)
1971	USA	Mariner 9—first Mars orbiter
1971	USSR	Mars 3—first Mars lander (lost contact after ~14s)
1971	USSR	Soyuz 11 disaster—air valve malfunction; cabin atmosphere bleeds off into space. Crew asphyxiates and dies just prior to re-entry.
1971	USA	Apollo 15—first manned lunar rover
1972	USA	Pioneer 10—first Jovian flyby
1972	USA	Apollo 17—last manned lunar mission to date
1973	USA	Skylab launched; first US space station
1973	USA	Pioneer 11—first Saturn flyby
1973	USA	Mariner 10—Venus flyby and first Mercury flyby

Year	Country	Event
1974	USSR	N-1 lunar launch vehicle suspended
1975	USA/USSR	Apollo-Soyuz mission
1975	USSR	Soyuz 18a suffers first manned launch abort en route to Salyut 4; almost lands in China. Instead lands in mountainous terrain and nearly falls off cliff. Crew suffers severe injuries; no fatalities.
1975	USA	Viking 1 successfully lands on Mars
1975	ESA	European Space Agency forms from ESRO
1976	USSR	N-1 lunar launch vehicle program cancelled
1976	USA	Enterprise Shuttle prototype rolled out
1977	USA	First Enterprise free flight
1977	USA	Voyager 2—Jupiter flyby, Saturn flyby, first Uranus flyby, first Neptune flyby
1979	ESA	Ariane-1 launches
1979	USA	Skylab deorbits
1980	USSR	Vostok explodes on pad, kills 48
1981	USA	First Shuttle flight STS-1, Columbia, orbital test flight
1982	USA	First operational Shuttle flight STS-5, Columbia
1983	USA	*Challenger* flies, STS-6

Year	Country	Event
1984	USA	*Discovery* flies, STS-41D
1984	USSR	Salyut-7—first EVA by a woman, cosmonaut Svetlana Savitskaya
1984	USA	*Challenger* STS-41B—first untethered EVA with MMU
1984	USA	*Challenger* STS-41G—first EVA by American woman, Kathryn Sullivan
1985	USA	*Atlantis* flies, STS-51J
1985	NASDA	Suisei and Sakigaki probes sent to Comet Halley
1985	USA	*Challenger*, STS-61A; first crew of 8
1986	USA	*Challenger*, STS-51L, explodes on ascent; all 7 crew lost
1986	USSR	Mir placed in orbit. First permanent manned space station.
1988	USSR	Buran, the Soviet version of a Space Shuttle, takes its only test flight, unmanned
1988	NASDA	HOPE program, a man-rated glide-return vehicle, funded
1988	USA	*Discovery* STS-26—Return To Flight
1988	ESA	Ariane-4 launches
1988	ESA/France	Mir—first EVA by French astronaut Jean-Loup Chrétien

Year	Country	Event
1989	USA	*Atlantis* STS-30—First shuttle to launch space probe—Magellan
1989	CSA	Canadian Space Agency formed
1989	USA	Galileo—first asteroid flyby, first Jupiter atmosphere probe
1990	USA	*Discovery* STS-31—Hubble Space Telescope released
1991	USSR/Russia	The USSR breaks apart
1992	USA	*Endeavor* flies, STS-49
1992	NASDA	*Endeavor* STS-47, Spacelab-Japan becomes the first entirely Japanese Space Shuttle payload
1993	China	Ministry of Aerospace Industry split into China National Space Administration and China Aerospace Corporation
1994	NASDA	First all-Japanese rocket, H-II, launched
1995	ESA/UK	*Discovery* STS-63—First EVA by Briton, Michael Foale; First EVA by a black African-American, Bernard A. Harris, Jr.
1996	USA	*Columbia* STS-80—longest Shuttle flight, 17 days 15 hrs
1996	ESA	First Ariane-5 launch fails (Ariane-5 intended to be man-rated)

Year	Country	Event
1996	China	First Long March 3B launch fails on liftoff; lands on nearby village
1997	USA	Mars Pathfinder lands via airbag system, releases Sojourner rover
1997	NASDA	*Columbia* STS-87—First EVA by Japanese astronaut, Takao Doi
1997	ESA	Second Ariane-5 launch is abort to orbit
1998	Russia	ISS Zarya module launched
1998	USA	*Endeavor* STS-88—first Shuttle-ISS mission delivers module Unity
1998	ESA	Third Ariane-5 launch successful
1998	JAXA	Nozomi Mars probe launched; failed to achieve Martian orbit
1999	China	Shenzhou-1 launched (unmanned)
1999	ESA	First commercial launch of Ariane-5
2000	Russia	ISS Zvezda module launched
2000	USA/Russia	ISS Expedition 1
2001	Australia	ISS—first EVA by Australian-born astronaut, Andy Thomas
2001	Canada	*Endeavor* STS-100—first EVA by a Canadian, Chris Hadfield
2001	Russia	Mir deorbits

Year	Country	Event
2003	USA	*Columbia* STS-107, disintegrates on reentry; all 7 crew lost
2003	USA	Mars Exploration Rover mission releases Spirit and Opportunity rovers
2003	China	Shenzhou-5 launched; first taikonaut
2003	JAXA	NASDA reorganized into JAXA
2003	JAXA	HOPE program canceled
2003	JAXA	Hayabusa-1 asteroid probe launch
2005	USA	*Discovery* STS-114—post-*Columbia* return to flight
2006	ESA/Sweden	*Discovery* STS-116—first Swede in space; first EVA by Swede, Christer Fuglesang
2007	JAXA	SELENE lunar orbiter launched (unmanned)
2008	China	Shenzhou-7—First taikonaut spacewalk, Zhai Zhigang
2008	China	Televised imagery of Chinese lunar rover
2009	JAXA	SELENE lunar orbiter impacts under control
2010	JAXA	Akatsuki Venusian probe launched; failed to enter Venusian orbit
2010	JAXA	IKAROS, the first solar sail, successfully deploys

Year	Country	Event
2011	USA	*Discovery* STS-133; last flight
2011	USA	*Endeavor* STS-134; last flight
2011	USA	*Atlantis* STS-135; last flight
2011	China	Launch of Tiangong-1, laboratory module and possible initial component of Chinese space station
2011	China	Shenzhou-8 rendezvous and docking with Tiangong-1; successful return to Earth
2011–12	USA	Voyager 1 passes heliopause
2013	China	Tiangong-2, second component of Chinese space station, planned launch
2014	China	Initiate unmanned Mars probes
2014	CSA	RADARSAT Earth obs Constellation initiates
2014	JAXA	Hayabusa-2 asteroid probe launch
2014	ESA/JAXA	BepiColombo probe launch to Mercury
2015	China	Tiangong-3, third component of Chinese space station, planned launch
2016	CSA	PCW polar obs Constellation initiates
2017	USA	SLS-1, unmanned lunar loop (Block I)
2018	Russia	New Vostochny Cosmodrome complete

Year	Country	Event
2019	USA	SLS-2, manned lunar loop (Block I)
2020-25	China	Complete Tiangong space station
2022	USA	SLS-3—manned launch (Block I)
2023	USA	SLS-4—manned launch (Block I)
2024	USA	SLS-5, first cargo configuration launch
2025	China	First taikonaut on Moon
2025	USA	SLS-6, "exploration" mission; destination as yet undefined (Block I)
2025	USA	Voyager 1 & 2 begin final power loss
2026	China	Initiate construction of lunar base
2026	USA	SLS-7—cargo launch (Block I)
2027	USA	SLS-8—manned mission; destination as yet undefined (Block I)
2028	USA	SLS-9—cargo mission (Block I)
2029	USA	SLS-10—manned mission; destination as yet undefined (Block I)
2030	USA	SLS-11—new cargo configuration (Block II)
2031	USA	SLS-12—manned mission; destination as yet undefined (Block II)

Year	Country	Event
2032	USA	SLS-13—new cargo configuration (Block II)
2040	China	First taikonaut (first human?) on Mars